Polymer Optical Fiber Bragg Gratings

Polymer Optical Fiber Bragg Gratings
Fabrication and Sensing Applications

Ricardo Oliveira, Lúcia Bilro, and
Rogério Nogueira

CRC Press
Taylor & Francis Group
Boca Raton London New York

CRC Press is an imprint of the
Taylor & Francis Group, an **informa** business

CRC Press
Taylor & Francis Group
6000 Broken Sound Parkway NW, Suite 300
Boca Raton, FL 33487-2742

First issued in paperback 2023

ISBN-13: 978-1-138-61262-4 (hbk)
ISBN-13: 978-1-03-265346-4 (pbk)
ISBN-13: 978-0-367-82270-5 (ebk)

DOI: 10.1201/9780367822705

Contents

Preface

Since the discovery of the Hill gratings in 1978, tremendous work has been carried out by many research teams in order to boost the technology of fiber Bragg gratings (FBG). Nowadays, FBGs can be inscribed transversely with a variety of methods, using ultraviolet, visible or femtosecond lasers, where the phase mask technique secures the most preferred choice.

Due to their ability to respond to external stimuli, the scientific community soon realized their benefits in the sensing area when compared with conventional electronic sensors. Among their inherent special characteristics are their small size, passive/low power, long-distance of operation, multiplexing capabilities, immunity to electromagnetic interference, high resolution, high signal-to-noise ratio and the ability to respond to a variety of parameters even in harsh environments. Because of those advantages, FBGs have matured to such a degree that they can now be found commercially available for a myriad of applications that span across areas such as civil engineering, medicine, gas, and oil industry, among many others.

Historically, the qualities of polymer optical fibers (POFs) have been overwhelmed by the popularity of the silica optical fibers mainly because of its higher losses. However, the improvement of the fiber fabrication technologies, as well as the use of low loss polymer materials, led these fibers to be considered a good opportunity for short-range communication and sensing applications.

Regarding specifically the sensing area, POFs offer a variety of opportunities over the well-known silica optical fiber, among them are their much higher: elongation; flexibility; failure strain; and thermo-optic coefficient. Furthermore, POFs have Young's modulus thirty times lower than silica fibers, making them very attractive for stress, force, pressure, and acoustic wave detection. Additionally, they also offer advantages for the monitoring of compliant structures, since the much stiffer silica optical fiber can reinforce the structure. Moreover, POFs can be made of different polymer materials allowing to tailor their final application, thus, they can be made for instance of hydrophilic or hydrophobic materials, giving the opportunity to develop humidity sensitive or insensitive sensors, respectively. Furthermore, contrary to silica optical fibers, POFs can be drawn under low temperatures, allowing the incorporation of organic materials without their destruction. POFs can also be considered biocompatible since they do not produce sharp edges when broken as silica fibers do, making them much more attractive.

The combination of POFs with the easiness of use of FBGs creates a symbiotic effect that benefits several sensing applications. This happens due to the qualities of polymers and the well-known advantages of FBGs. The combination of these two technologies started in 1999 with the demonstration of the possibility to inscribe gratings in both multimode and single-mode fibers. The technology has then been used for the production of sensors with interesting characteristics. The advent of microstructured polymer fibers brought another opportunity for the polymer optical fiber Bragg grating (POFBG) technology, due to the ease in achieving single-mode behavior. Also, the incorporation of different dopants and the use of different laser sources allowed several breakthroughs along the years, in such way that the technology is poised to make the leap from the laboratory to the market.

This book gives an overview of the POFBG technology over the last 20 years, covering aspects related to the fiber Bragg grating fabrication and also sensing applications. This book is split into five chapters and it is written in such a way that can provide a comprehensive and simple route to new users, scientists and engineers working or wishing to work, in the field of POFBGs.

The first chapter covers the main fundamentals of POFs, their material properties and the technologies commonly used to get the POF ready to work, such as the fiber end face termination and their connectorization or splicing process.

Chapter 2 brings an overview of the fabrication technologies used for the inscription of Bragg gratings in POF. As it occurs with silica optical fibers, the photosensitivity in polymers is also a hot topic, and different wavelength sources, materials and dopants have been explored along the years and thus, they are covered in this chapter.

POFBGs have matured throughout the years and because of that the technology is not only restricted to uniform gratings. Instead, special grating profiles have been implemented through different technologies, allowing to spread their range of applications. This motivated the writing of Chapter 3, which will be dedicated to a more specialized type of POFBGs.

Chapters 4 and 5 are intended to show the different applications of POFBGs for sensing applications. Therefore, Chapter 4 covers the applications of POFBGs in general terms, exploring the POFBG capability to measure properties such as strain, temperature, water absorption, or hydrostatic pressure. On the other hand, Chapter 5 covers applications of POFBGs in more specific cases, as is the case when POFBGs are embedded or attached in special arrangements, materials and a combination of both. Such configurations allow an easier interaction of the external environment with the POFBG, allowing the opportunity to explore a variety of sensing areas, mainly because of its higher flexibility, elasticity, and low Young's modulus. Nevertheless, the multiparameter opportunities of POFBGs also bring cross-sensitivity issues when single parameter is required. Hence, special fiber optic arrangements are also under the scope of this chapter.

To conclude, it is our hope that this book will be a useful reference to the reader regarding the fabrication and application of polymer optical fiber Bragg gratings.

We would like to thank the support given along the last ten years by FCT-Fundação para a Ciência e Tecnologia through Portuguese national funds, for the founding support on the subject area of this book, namely through research projects such as POFcom (PTDC/EEA-TEL/122792/2010); hiPOF (PTDC/EEI-TEL/7134/2014); AQUATICsens (POCI-01-0145-FEDER-032057); investigator grant IF/01664/2014; project INITIATE and UID/EEA/50008/2019.

Ricardo Oliveira
Lúcia Bilro
Rogério Nogueira
Instituto de Telecomunicações, Portugal

Authors

Ricardo Oliveira received his Ph.D. degree in Physics Engineering from the University of Aveiro, Portugal in 2017. He is now a researcher at Instituto de Telecomunicações, Aveiro, Portugal. Ricardo has been working on several research projects involving fiber optic devices and components. His main interests include optical fiber devices and components for communications and sensors using polymer optical fibers, fiber Bragg gratings, and fiber optic interferometers.

Lúcia Bilro is presently a researcher at Instituto de Telecomunicações (IT) in Aveiro-Portugal. She is a co-founder and CTO of Watgrid, Lda and board member of the Portuguese Society of Optics and Photonics (SPOF). Prior she was a post-doc research fellow at IT and Invited Professor at Polytechnic Institute of Viana do Castelo. She received her Ph.D. degree in Physics from the University of Aveiro, Portugal in 2011. Her current research interests are Bragg sensors, plastic optical fiber sensors, physics, medicine, rehabilitation, and environmental monitoring.

Rogério Nogueira is a principal research scientist at Instituto de Telecomunicações (IT) in Aveiro-Portugal, where he coordinates the group Optical Components and Sub-systems. He is a co-founder and CEO of Watgrid, Lda. Prior, he was an invited professor at the University of Aveiro and Innovation Manager for both Coriant and Nokia Siemens Networks Portuguese branches. Dr. Nogueira is co-founder and vice-president of the Portuguese Optical Society. He received his Ph.D. degree in Physics from the University of Aveiro in 2005 and an Executive Certificate in Management and Leadership from MIT in 2011. His current research interests are Optical Communications, Fiber Optical Components, Fiber Sensors, and Microwave Photonics.

1 Principles
POF Materials, Termination and Connectorization

1.1 INTRODUCTION

Nowadays, the unique properties of polymer optical fibers (POFs) have fostered the research on the development of different components, systems, and sub-systems, most of those targeting telecommunications and sensing applications. When compared with the well-known silica fiber, POFs can bring new features, such as high elasticity and flexibility, a thirty times lower Young's modulus, a negative and much higher thermo-optic coefficient, lightweight, non-brittle nature, the capability to absorb water, biological compatibility, among others. POFs can be made by one or more materials and thus specific applications may be tailored by the use of a particular polymer material. Nevertheless, there is a variety of transparent polymer materials, each one presenting its own physical and chemical properties. The possibility of doping POFs with organic materials, including fluorescent or photosensitive materials is also feasible, opening a plethora of new applications. Nowadays, POFs are available with a variety of refractive index profiles, materials and also with special hole arrangements in the cross-sectional area that run along the length of the fiber, commonly referred to as microstructured POFs (mPOFs). These characteristics allow better control of light guiding mechanisms. Because of that, POFs can now be easily fabricated with single-mode (SM) behavior in a wide range of wavelengths. Thus, POF sensors are not only limited to intensity-based techniques but can also be used in interferometric, polarimetric, and wavelength-based schemes, opening new sensing opportunities. However, the requirement for SM behavior leads to having fibers with core diameters of few micrometers. Consequently, conventional cleaving methods using sharp razor blades cannot be used, since the defects created by the blade are of the same order of the diameter of the POFs core. Hence, the research community has been actively involved in solving the problem and a variety of methodologies capable to produce reliable POF end face terminations have been proposed. Another problem faced by small core fibers is related with the splicing process, namely to the silica optical fiber, which is normally used as the light delivering system due to the availability of devices and components already developed for the telecom industry. The different melting point of these two fibers makes the conventional hot melting process infeasible. Therefore, different techniques have been used, such as the ferrule connectorization process and the butt coupling process using UV curing adhesives.

This chapter will start by introducing the special characteristics of different polymer optical materials, giving special focus to their physical and chemical properties.

Then, the end face termination of POFs through different techniques will be reviewed for the most attractive technologies. Finally, different processes used to couple light into small core POFs will be discussed, giving emphasis to the butt coupling through UV curing adhesives.

1.2   POLYMER OPTICAL FIBERS

In the early 1950s, some of the first optical fibers were made of polymer core surrounded by beeswax, acting as the cladding material. Later on, the coating material was replaced by a liquid polymer that was cured and finally painted with black lacquer to prevent leakage of light [1]. Optical fiber production processes have then evolved and in 1963, DuPont manufactured the first POF fiber, Crofon™ [2,3]. The fiber had a step-index (SI) profile (see Figure 1.1), and had a multimode core composed of polymethylmethacrylate (PMMA) material. Fifteen years later, Mitsubishi Rayon Co., Ltd., introduced the first commercial POF (Eska™), composed of a SI profile and made of PMMA [4,5]

SI fibers are designed to have a constant refractive index for the core and cladding, as represented at the bottom right-hand side of Figure 1.1.

The occurrence of SM behavior is only possible when the normalized frequency defined as:

$$V = \frac{2\pi a}{\lambda} NA \tag{1.1}$$

is lower than 2.405 [6], where a is the core radius, λ is the free space wavelength and NA is the numerical aperture described as:

$$NA = \sqrt{n_{co}^2 - n_{cl}^2} \tag{1.2}$$

where n_{co} and n_{cl}, are the core and cladding refractive indices of the fiber, respectively. When multimode (MM) behavior is observed, the number of modes (N) propagating in an optical fiber is according to the electromagnetic theory, given by the following expression:

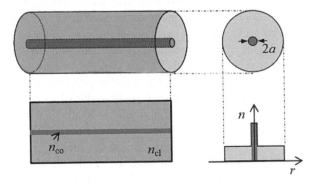

FIGURE 1.1   Schematic representation of a SI fiber in different projections. The refractive index profile is also shown at the bottom right-hand side.

$$N \approx \frac{1}{2} \frac{g}{g+2} V^2 \qquad (1.3)$$

where $g = 2$ is found for fibers with GI profile and in the limit ($g \to \infty$), for a SI fiber, N becomes:

$$N = \frac{V^2}{2} \qquad (1.4)$$

As an example, a PMMA fiber operating at its low-loss region of 650 nm, with $a = 0.5$ mm and $NA = 0.5$, will exhibit 2.9 million modes.

Fibers with SM behavior are manufactured in such a way that the core diameter and the refractive index of the core and cladding regions, only allow the propagation of one mode at a given wavelength. A mode, using the waveguide theory, is a spatial distribution of optical energy in one or more dimensions that remains constant in time. This type of fiber is widely used in telecommunications due to the increased bandwidth associated with the absence of pulse broadening along the fiber length as it occurs with MM fibers. Regarding sensing applications, specifically the ones based on fiber Bragg gratings (FBGs), SM fibers are the most preferred choice due to the single peak spectrum that allows to obtain better sensing resolutions. Furthermore, the multi-peak spectrum of a MM-FBG has inherent problems associated with the power exchange between modes, making it difficult to track the FBG wavelength response.

Transparent polymer materials can be found in a wide range of refractive indices, which vary from 1.32 for highly fluorinated acrylic to around 1.6 for some cast phenolic resins, and because of that, a high refractive index contrast between the core and cladding can be achieved, allowing to have fibers with high NA, which makes them very efficient for collecting light. However, to achieve SM behavior, a balance between the core radius and the refractive index contrast needs to be found. While one could think in using different materials for the core and cladding materials (high refractive index contrast), this would require the use of thin cores, which would lead to have higher scattering losses at the core cladding interface. According to Equation 1.1, the scenario becomes even worse if one thinks to operate the fibers in the low loss region of polymers since it would require an even smaller core. The solution was to use dopants in both core and cladding materials. However, the dopant diffusion needs to be carefully controlled in order to have a stable refractive index, making the task very challenging.

Silica SM fibers have been developed since the early sixties, but the higher attenuation presented by polymers delayed the SM-POF demonstration to the early nineties by Kuzyk and co-workers [7]. The fiber was made of PMMA and the core was dye-doped with Squarylium dyes, Disperse red 1 Azo dye and Phthalocyanine dyes. The resulting fiber after drawing had a core and cladding diameter of 8 and 125 µm, respectively, with an attenuation coefficient of 0.2–0.6 dB/cm at the operational wavelength of 1,300 nm [7]. Later, Bosc and Toinen reported a SM-POF in which the refractive index contrast between core and cladding was adjusted using copolymers instead of dyes for both core and cladding [8]. The core was composed of copolymers methyl methacrylate (MMA) and ethyl methacrylate (EMA), while the cladding was composed of copolymers MMA and trifluoroethyl methacrylate (TFMA), where the fiber diameter was roughly estimated to have 145 μm and the core ~5.8 μm, being the losses measured at 850 and 1,550 nm equal to, 0.05 and 0.20 dB/cm, respectively. The authors also propose the co-polymerization of MMA with either styrene or TFMA, in order to reach a large range of core refractive index from 1.42 to 1.59 [8]. The commercialization of this type of fibers was later done by Paradigm Optics, Inc., being the fibers made of PMMA with cores copolymerized with polystyrene (PS) (<3%). The fibers were sold with trade names of SM-MORPOF02 and SM-MORPOF03 with a cut-off wavelength around 1,100 and 750 nm, respectively. However, SM SI-POFs based on other materials have also been reported, such as the use of TOPAS® 5013S-04 core and ZEONEX® 480R cladding [9] and also POFs made of the most transparent polymer material, composed of a perfluorinated graded-index (GI) core and PMMA overcladding [10].

The creation of SI-POFs may be done through two processes, continuous or discontinuous, being those related to the type of flow. Regarding the latter, it comprises a two-step procedure that involves preform fabrication and the drawing process. The preform, which is a scaled-up (larger in diameter, shorter in length) version of the fiber, is heat-drawn using a drawing tower. The preform fabrication may be done: (i) by filling a PMMA hollow cylinder with a mixture of monomers in different proportions [7]; (ii) through the "Teflon Technique" [11], in which the cladding material is first polymerized in a glass tube containing a Teflon rod in the center, that is removed in a second step, in order to be filled with the core monomers; (iii) through the "pull-through" technique [12], in which the cladding preform is made as it is done for the "Teflon Technique," but the core preform is polymerized in a tube that will later be heat drawn and tightly inserted in the hollow cladding preform; and also (iv) through heat casting [9], in which the cladding material granulates are cast into a solid rod that is later drilled in the center for the injection of the molten polymer core. Discontinuous processes are thus limited to the length and diameter of the preform, leading to a disruption of the drawing process, which inherently increases the fabrication costs. However, in continuous manufacturing techniques, the procedures involved are made simultaneously, allowing the fabrication of a fiber with a theoretical infinite length. Overall, the main advantage when compared with discontinuous manufacturing techniques is the high production rate that can be obtained, thus providing better economies.

Generally, discontinuous processes have been the most reported by the research community, for the manufacture of SM SI-POFs, specifically the ones involving polymerization process [7,13–16], since it allows fine control of the refractive index between the core and cladding. This control may be done through the introduction of monomers and dopants in different proportions for the core and cladding. Common examples of monomers are: MMA, benzyl methacrylate (BzMA), EMA, TFMA [8,16], butyl methacrylate (BMA) and dopants such as rhodamine [11], fluorescein [17], trans-4-stilbenemethanol (TSB) [14], benzildimethylketal (BDK) [18], methyl vinyl ketone (MVK) [19], 9-vinylanthracene (9-VA) [20], benzophenone (BP) [15], diphenyl sulfide (DPS) [21,22], diphenyl disulphide (DPDS) [12], etc. The role of each monomer can be classified as:

- MMA to act as the main monomer;
- BzMA to increase the core refractive index and also to enhance the UV absorption due to the phenyl group (pointed as the main reason for the photosensitivity in poly(MMA-co-BzMA) POFs) [15];
- TFMA a low index fluorinated monomer used to reduce the cladding refractive index;
- BMA, used mostly in the cladding to lower the heat drawing temperature of the preform (mostly to avoid the decomposition of dopants [23]);
- and dopants for enhanced photosensitivity and also to increase the core refractive index as is the case of DPS and DPDS [12,21].

The refractive index of a compound may be estimated from its molar refraction and molecular volume according to the Lorentz–Lorentz equation [24]:

$$\frac{n^2-1}{n^2+2} = \frac{[r]}{\upsilon} \equiv \varphi \qquad (1.5)$$

$$n = \sqrt{(2\varphi+1)/(1-\varphi)} \qquad (1.6)$$

where n is the refractive index, $[r]$ is the molar refraction, υ is the molecular volume and φ is the molar refraction per unit volume. Therefore, the refractive index of the fiber core ($j=1$) and cladding ($j=2$) can be given by:

$$\varphi_j = \sum_{i=1}^{N} c_i \frac{n_i^2-1}{n_i^2+2} \qquad (1.7)$$

$$n_j = \sqrt{(2\varphi_j+1)/(1-\varphi_j)} \qquad (1.8)$$

where c_i and n_i are the volume fraction and refractive index, respectively, of unit i, in which $i=1$ stands for instance to MMA, $i=2$ to EMA, and so on, for all the materials used for the polymerization reaction in the core and cladding.

1.2.1 POLYMER OPTICAL FIBER MATERIALS

To attend the specifications of a given sensing application, an optical fiber sensor needs to be designed carefully. The type of the fiber (e.g., SI, GI, microstructured), the fiber material (e.g., polymer, silica, hybrid), its transparency and the mechanical, thermal and chemical properties are key parameters that have to be considered.

Polymer optical fibers have been pointed out as a viable alternative to silica-based fibers. The reason is mostly due to the unique properties of polymers, which can be attributed to the lower Young's modulus, higher elastic limit, negative and much higher thermo-optic coefficient, flexibility, biocompatibility, non-brittle, among others. Furthermore, POFs have the possibility to be doped with organic materials, including fluorescent or photosensitive materials such as

Rhodamine dyes [25–27]. These can be added during the preform polymerization or can be diffused into the polymer matrix through solution doping [26,28].
This particular characteristic cannot be achieved with silica fibers since the
drawing temperature is much higher than the one for which all organic materials
decompose (~400°C) [29].

Polymers are available in a wide range of materials; however, for the fabrication of an optical fiber, transparency is the most required property. Among the
different transparent polymer optical fiber materials used for the fabrication of
POFs, the most used are: PMMA; poly(ethyl methacrylate) (PEMA); poly(benzyl
methacrylate) (PBzMA); poly(butyl acrylate) (PBA); poly(trifluoroethyl methacrylate) (PTFMA); polycarbonate (PC); cycloolefin polymers (COP), e.g., ZEONEX®
480R; cycloolefin copolymers (COC), e.g., TOPAS® 5013L; polystyrene (PS); and
perfluorinated polymer (PF), e.g., CYTOP®. Chemical structures of these polymer
materials may be seen in Figure 1.2 and their most important properties are shown
in Table 1.1.

The wide range of values presented by different polymer materials for each of the
properties shown in Table 1.1, makes the development of a specific POF application
very appealing. Thus, a variety of fields can benefit from these properties, such as
in short-range transmission systems [41], sensing applications (e.g., biosensing [42],
civil engineering [43], medical [44], etc.), and lighting [45]. Nevertheless, among
the polymer materials, there are different features that may benefit some particular

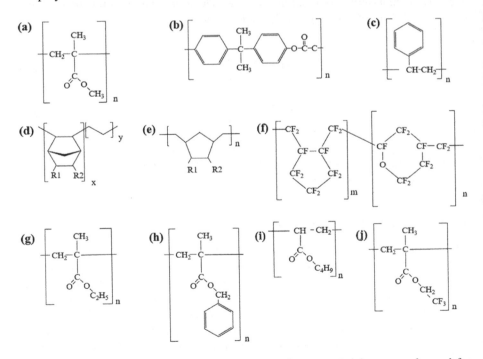

FIGURE 1.2 Chemical structures of transparent polymer materials commonly used for
the production of POFs, specifically: PMMA (a); PC (b); PS (c); COC (d); COP (e); PF (f);
PEMA (g); PBzMA (h); PBA (i); PTFMA (j).

TABLE 1.1

Properties of Materials Commonly Used for the Fabrication of POFs

Material	Refractive Index (n_D)	Density (g/cm³)	Abbe Number	Water Absorption (%)	Young's Modulus (GPa)	Yield Strength (MPa)	Yield Strain (%)	Differential Stress–Optic Coefficient (×10⁻¹²/Pa)	Poisson Ratio	Glass Transition Temperature (°C)	Thermo-Optic Coefficient (×10⁻⁵/°C)	Thermal Expansion Coefficient (×10⁻⁵/°C)
PMMA	1.49 [30]	1.19 [30]	57 [31]	0.3 [30]	3.3 [30]	77 [30]	2.3ᵃ	−6.0 [32]	0.40 [33]	107 [30]	−9 [32]	6 [30]
PC	1.59 [34]	1.20 [34]	30 [32]	0.2 [34]	2.5 [34]	63 [34]	2.5ᵃ	72.0 [32]	0.37 [33]	143 [34]	−14 [32]	7 [34]
PS	1.59 [35]	1.04 [35]	31 [31]	0.2 [31]	2.9 [35]	45 [35]	1.6ᵃ	−55.0 [32]	0.35 [33]	99 [35]	−12 [32]	9 [35]
TOPAS® 5013	1.53 [36]	1.02 [36]	56 [36]	<0.01 [36]	3.2 [36]	46 [36]	1.4ᵃ	−4.0 [36]	0.37 [36]	134 [36]	−10 [32]	6 [36]
ZEONEX® 480R	1.53 [37]	1.01 [37]	56 [37]	<0.01 [37]	2.2 [37]	59 [37]	2.7ᵃ	6.5 [32]	0.40 [37]	138 [37]	−13 [32]	7 [37]
CYTOP®	1.34 [38]	2.03 [38]	90 [38]	<0.01 [38]	1.5 [38]	40 [38]	2.7ᵃ	6.5 [38]	0.42 [38]	108 [38]	−5 [39]	12 [38]
PEMA	-	1.49 [40]	-	0.2 [40]	-	-	-	-	-	63 [40]	−11 [40]	6 [40]
PBzMA	-	1.57 [40]	-	-	-	-	-	4.0 [40]	-	54 [40]	-	17 [40]
PBA	-	1.47 [40]	-	-	-	-	-	−0.5 [40]	-	53 [40]	-	-
PTFMA	-	1.42 [40]	-	-	-	-	-	-	-	69 [40]	-	-

ᵃ Calculated from the stress-strain relation, using the yield strength and Young's modulus.

application. For instance, the heat and chemical resistance or water absorption are different from polymer to polymer. Those characteristics can be considered as an advantage or drawback depending on the desired application.

1.2.1.1 Polymethylmethacrylate (PMMA)

The most well-known and widely used polymer optical fiber material is PMMA. The reasons are essentially related to its high qualities, such as good transparency at the visible region, relatively low cost, ease of processing, high yield strain, etc. PMMA is fabricated through the polymerization of the monomer MMA, forming long chains with a typical molecular weight of around ~10^5 [46]. Furthermore, PMMA can be altered by copolymerization with other acrylic monomers such as EMA, BzMA, etc., allowing to control the refractive index or the glass transition temperature (T_g) [16], etc., which is important to tailor specific applications. Nevertheless, the possibility of doping PMMA fibers allows the development of fiber-based devices, such as lasers and amplifiers [27], and brings also the opportunity to enhance the photosensitivity for effective fiber Bragg grating inscription [14,18].

As may be seen from Figure 1.2a, PMMA is composed of eight aliphatic C–H bonds, being the overtones of the C–H vibration, the main reason for the losses presented by this polymer at the infrared region [47]. PMMA is resistant to most alcohols and acids [46], making their use in such conditions possible. Furthermore, PMMA has an affinity to water and because of this particularity, POFs based on this material have been proposed for relative humidity (RH) detection [48], or even to detect concentration in solutions [49]. Nevertheless, this property is also problematic in other sensing applications, due to the cross-sensitivity issues, as well as the long-term signal stabilization, related to the water absorption by the polymer matrix.

1.2.1.2 Polycarbonate (PC)

Polycarbonate presents superior performance over PMMA regarding its mechanical properties and high operation temperature, (see Table 1.1). Its refractive index is about 1.59 (@ 589 nm) [28], which is higher than most of the transparent polymer materials. This characteristic is thus motivation for its use as the core material in POFs. However, if compared with PMMA, PC has an attenuation four times higher, hence, its use is limited to short-range applications. Examples may be found in the automotive engineering, specifically in the engine area, where temperatures can go up to 125°C [46], which is still below the glass transition temperature of PC (T_g = 143°C [34]). Furthermore, PC can also absorb water from the environment, allowing the development of fiber optic RH sensors. However, it tends to degrade at high-temperature and humid environments [46].

1.2.1.3 Polystyrene (PS)

Polystyrene is a costless transparent material that can be used for the fabrication of POFs, namely as the core material, due to its higher refractive index (1.59 @ 589 nm [35]). Despite its lower mechanical and thermal properties, when compared with PMMA, PS shows the possibility to be thermally polymerized without any initiator as is needed in PMMA, which sometimes produces bubbles during the polymerization reaction [50].

Regarding attenuation, PS has lower transparency than PMMA. The reason may be attributed to the overtones of the three aliphatic and five aromatic carbon-hydrogen (C–H) bonds [51], (see Figure 1.2c), and also due to the phenyl group that is present in each monomer unit, which due to the flat physical geometry gives rise to molecular anisotropy and hence, scattering [52]. Because of that, side-emitting or fluorescent fibers made of PS are being used [53]. Compared with PC, PS shows lower attenuation, which can be attributed to the number of phenyl groups present in each monomer (two for PC and just one for PS, as may be seen in Figures 1.2b and c).

1.2.1.4 Cycloolefin Copolymer (COC) and Cycloolefin Polymer (COP)

The randomness of molecular chains is essential to reach polymer transparency. The refractive index of crystalline and amorphous regions in a polymer material is different and thus light passing between those regions will be scattered, causing haziness and thus low transparency. Polypropylene (PP) and Polyethylene (PE) are popular polyolefin polymers used in many industrial applications. However, their use in optical components is limited due to their low transparency, associated with the light scattering at the interface between the crystalline and amorphous parts in the polymer matrix [54].

However, COCs and COPs are a new class of thermoplastics, which are essentially amorphous, having excellent optical properties as a result of the presence of cyclic structures in their polymer chain (see Figures 1.2d and e). Among the major characteristics, it may be found: low specific gravity; high durability; good transparency; low water absorption; low scattering; low birefringence; heat resistance; high flowability, etc.

COCs can be processed through copolymerization of cycloolefin, such as norbornene or cyclopentene, with ethylene or alpha-olefin, using a metallocene catalyst. They have been commercialized with names: APEL™ by Mitsui Chemicals Co., Ltd and TOPAS® by Topas Advanced Polymer GmbH.

In COPs, the polymerization is made through Ring Opening Metathesis Polymerization (ROMP) of norbornene derivatives followed by hydrogenation of double bonds, conferring more stability in terms of heat and weather resistance [54,55]. COPs are sold with trade names of ZEONEX® and ZEONOR® by Zeon, and ARTON® by Japan Synthetic Rubber [32].

Due to the special characteristics of COCs and COPs, they have been reported for the fabrication of POFs, namely, microstructured [56–60], SI [9,61] and multicore composite SM fibers [62]. Their transparency allowed the utilization in different wavelength regions, such as near-infrared [9,56,58,59,61], and terahertz (THz) [57]. Furthermore, other properties such as biocompatibility [56], low moisture absorption [9,59,61], high-temperature resistance [9,58], have also motivated the development of different fiber optic applications.

1.2.1.5 Perfluorinated (PF) Polymer

Perfluorinated polymers are formed via a free radical mechanism of perfluoromonomers. They tend to form partially crystalline structures, and thus they are opaque due to the light scattering between the amorphous and crystalline phases. The most efficient way to give transparency to the polymer is through the introduction of aliphatic rings into the main chain, blocking the formation of crystalline structures. An example of such amorphous polymer is CYTOP®, developed by Asahi Glass Co., Ltd.

(AGC) [51]. CYTOP® is a homopolymer obtained from perfluoro(4-vinyloxyl-1-butene), yielding penta- and hexa-cyclic structures in the polymer chain [63] (see the structures shown in Figure 1.2f). Because of that, CYTOP® presents the highest transparency amongst the optical fiber materials. In view of that, POFs based on this material have been developed and commercialized, namely Lucina® fiber, fabricated by AGC. The fiber was MM and had a GI profile, having losses of 10 and 15 dB/km, at 1,000 and 1,300 nm, respectively [64]. Nevertheless, the theoretical loss limit for those regions is estimated to be around 0.7 and 0.26 dB/km, respectively [51,65], a value that could be compared with that of silica fibers. These outstanding low losses are possible due to the absence of hydrogen atoms in the main chain, contrary to most POF-based materials. Thus, the molecular absorption is only due to the overtones of C–C, C–F, and C–O bonds, which are lower in intensity, since the wavelengths of their fundamental stretching-vibration are found at longer wavelengths [51].

Additionally to the outstanding transparency, fluorinated polymer fibers also have thermal resistance, chemical durability, low water absorption, and low refractive index, etc. [38,63], making them very attractive for the development of almost any type of POF sensor.

1.2.2 REFRACTIVE INDEX

The refractive index of a material is a non-dimensional number that describes how fast the light can propagate through a medium. It is defined as:

$$n = \frac{c}{v} \tag{1.9}$$

where c is the speed of light in vacuum and v is the phase velocity of light in the medium. The variation of the refractive index depends on the molecular polarizability and density as described by the Lorentz–Lorentz equation. Nevertheless, the refractive index of materials varies with the wavelength and it is defined as dispersion. The amount of dispersion can be quantified in the wavelength region where the human eye is most sensitive by the Abbe number (V_D), described as:

$$V_D = \frac{n_D - 1}{n_F - n_C} \tag{1.10}$$

where, n_D, n_F, and n_C are the refractive indices of the material at the wavelengths of the Fraunhofer D-, F- and C-lines (589.3, 486.1, and 656.3 nm, respectively). Materials with high refractive index, such as PC, will present a low Abbe number (see Table 1.1), and therefore, higher dispersion. However, materials such as CYTOP® have the lowest refractive index among the optical plastics and thus it presents the highest Abbe number (see Table 1.1) and the lowest dispersion. The refractive index–wavelength dependence for wavelengths other than the spectral lines can be calculated through a more accurate method, over the total transmission region, by well-known Sellmeier empirical formula:

$$n^2(\lambda) = 1 + \frac{B_1\lambda^2}{\lambda^2 - C_1} + \frac{B_2\lambda^2}{\lambda^2 - C_2} + \frac{B_3\lambda^2}{\lambda^2 - C_3} \qquad (1.11)$$

where B_1, B_2, B_3, C_1, C_2, and C_3 are the Sellmeier coefficients that are frequently quoted instead of refractive index in tables. For comparison purposes, the refractive index wavelength dependence of different transparent polymer materials commonly used for the production of POFs is shown in Figure 1.3.

From Figure 1.3, it can be seen that materials with high refractive index, such as PC or PS, present strong dependence with wavelength, as compared with materials with low refractive index, such as CYTOP® and thus higher material dispersion is expected for the first.

Transparency is one of the most required properties for the design of an optical fiber. It is dependent on: the molecular structure; the molecular conformation; and the presence of impurities. Transparency is also related to the refractive index since reflectance (R_0) depends on the refractive index as follows [32]:

$$R_0 = \frac{(n_1 - n_2)^2}{(n_1 + n_2)^2} \qquad (1.12)$$

where n_1 and n_2 define the first and second medium where light is being transmitted (considering normal incidence). Thus, the most transparent materials are the ones that present a lower refractive index. Hence, it is not surprising to find CYTOP® as the most transparent polymer optical fiber material, and thus the preferred choice for POFs in optical communications and distributed sensor systems.

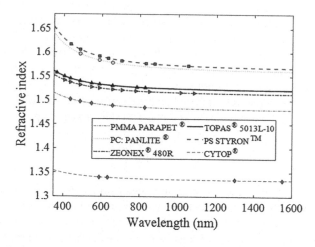

FIGURE 1.3 Refractive index evolution of common transparent polymer materials used for the production of POFs. The marker points were obtained from experimental values found in the literature for PMMA, PC, ZEONEX®, TOPAS®, PS, and CYTOP® [32,36,38,66]. The curves are the corresponding fittings, following the Sellmeier dispersion formula shown in Equation 1.11.

1.2.3 Biocompatibility

Nowadays due to the biocompatible nature of polymers, POFs have been proposed for in-vivo applications. When compared with the well-known silica fiber, which is also considered biocompatible, POFs are intrinsically safer. This occurs due to the brittle nature of silica fiber which can break and have a high chance to create a sharp edge that could damage the surrounding tissue, a problem that the much softer POF would never pose. Nevertheless, POFs can be modified to detect specific biochemical species [56,67] and can be also made to be biodegradable and biocompatible [68]. Furthermore, the development of polymer optical fiber Bragg gratings (POFBGs), brought new sensing opportunities due to the easy integration and characterization. Therefore, an extensive set of applications have been demonstrated, such as blood pressure monitoring [69], pressure in endoscopic applications [70], optoacoustic endoscopy [71–73], heartbeat monitoring [12], respiratory monitoring [12,69,74], erythrocyte detection [75], foot plantar pressure [76] and also human-robot interaction forces [77].

1.2.4 Attenuation in POFs

The attenuation mechanisms in a POF can be grouped as extrinsic and intrinsic. Regarding the extrinsic attenuation (α_i), it is due to the absorption by organic contaminants, transition metals and water within the fiber material. It is also due to the scattering of light in micro-voids and dust as well as core diameter fluctuations, birefringence and core cladding imperfections during fiber manufacturing process. Ultimately, these are the losses that can be reduced by purification of the materials and by the optimization of the manufacturing process [78]. On the other hand, intrinsic attenuation is the result of the physical and chemical structure of the fiber material, which is due to scattering and absorption, namely: the Rayleigh scattering loss (α_{Rs}), the electronic transition absorption (α_{eT}), and the molecular vibration absorption (α_{mv}). These contributions depend on the composition of the optical fiber and cannot be eliminated, representing thus the ultimate transmission loss limit. α_{Rs} is caused by inhomogeneities of random nature and it is inversely proportional to the 4th power of the wavelength. α_{eT} is responsible for absorptions in the UV region associated with the light-induced electronic transitions in polymers. The peak absorption tails to the visible region and obeys to the Urbach empirical rule [79]:

$$\alpha_{eT} = A_0 \exp\left(\frac{B_0}{\lambda}\right) \tag{1.13}$$

where A_0 and B_0 are the material constants determined from the absorbance spectrum. For PMMA, the loss is almost negligible after 450 nm, which is explained by the weak absorption at 220–230 nm of the $\eta \to \pi^*$ transition of the double bond of the ester group, and $\pi \to \pi^*$ transitions of low intensity near 200 nm due to the azo group in a polymerization initiator (when azo compounds are used) [80,81]. However, if polymer materials, such as PS or PC are used as the main host material of a POF, losses as high as 100 and 750 dB/km at 500 nm for PS and PC, can be achieved, respectively. The explanation for those higher losses are the result of the $\pi \to \pi^*$ electronic transitions of the phenyl groups [80]. Both α_{Rs} and α_{eT} have little influence on the attenuation loss in wavelength regions longer than 700 nm. However, α_{mv} plays a predominant role in the fiber attenuation at the infrared and visible spectral regions. This occurs because the basic constituents of polymers are made of carbon-hydrogen bonds which have their fundamental molecular vibration absorption localized at ~3,500 nm with overtones that spread up to the visible region, being the strength of each overtone decreased one order as the vibrational quantum number increases by one [47,78]. Since the fundamental vibration absorption occurs at longer wavelengths, the occurrence of high attenuation at the infrared region is obvious, when compared to the visible region. One possible way to solve this could be through the substitution of hydrogen atoms with heavier elements, such as deuterium or fluorine, allowing to shift the excitation of vibrations towards higher wavelengths, reducing the loss and increasing the usable range of wavelengths. However, the production of such materials is difficult to synthesize and it is costly. Therefore, these materials are not the most used for the fabrication of POFs.

Considering that the state of art of the POF manufacturing process (namely SM-POFs) is essentially found at research labs and universities, both intrinsic and extrinsic losses can pose challenges on the use of POFs, mainly at the infrared region where material losses are extremely high. It is not surprising to find SM-POFs either SI and mPOFs, presenting losses at the 1,550 nm region, as high as 1–3 dB/cm [82–84]. Consequently, the application of these fibers at those wavelengths is limited to a few centimeters. To summarize, the losses at different spectral regions for different transparent optical materials used to fabricate POFs, may be seen in Figure 1.4 (data collected from different authors). For comparison purposes, the same figure also contains data related to the silica material.

As can be seen in Figure 1.4, CYTOP® and PMMA-d8 materials present the best loss performances and thus they are the best candidates for the development of low loss POF. Despite the higher loss presented in the infrared region, it is still possible to develop applications in this spectral region. Examples are point sensors, such as FBGs, which normally require fiber lengths of few centimeters [91,92]. However, for the development of POF distributed or quasi-distributed sensors, the use of the visible spectral region and/or low loss polymers is mandatory. One candidate for such applications is CYTOP®, which presents low loss at the visible and infrared regions, being the attractiveness of the later infrared region, well-known due to the possibility to use the already developed silica optical fiber-based devices.

1.2.5 POFs Mechanical Properties

One of the great advantages of POFs when compared to silica optical fibers, is related to their mechanical performance. Because of that, they have been considered as a good opportunity for strain sensing applications. Nevertheless, the implementation of special sensing schemes in which the variable to be measured is converted into strain in the POF, may also be used to measure other quantities, such as pH [93], liquid level [94], curvature [95], heartbeat [12], respiration, blood pressure [69], etc. It is thus important to know, how the fiber will respond when subjected to strain and also the strain and stress limits. One common method used to get such knowledge is

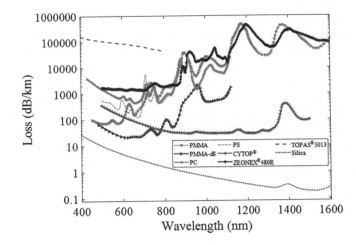

FIGURE 1.4 Loss spectrum of different transparent optical fiber materials, namely PMMA, PMMA-d8, PC, PS, CYTOP®, ZEONEX® 480R, TOPAS® 5013 and silica [85–90]. (Adapted from G. Khanarian and H. Celanese, "Optical properties of cyclic olefin copolymers," *Opt. Eng.*, vol. 40, no. 6, pp. 1024–1029, 2001; A. Argyros, "Microstructured polymer optical fibers," *J. Light. Technol.*, vol. 27, no. 11, pp. 1571–1579, 2009; M. Naritomi, "'CYTOP®' amorphous fluoropolymers for low loss POF," in *POF Asia-Pacific Forum*, 1996, pp. 23–27; T. Yamashita and K. Kamada, "Intrinsic transmission loss of polycarbonate core optical fiber," *Jpn. J. Appl. Phys.*, vol. 32, no. 6A, pp. 2681–2686, 1993; "Paradigm the Polymer Division of Incom, Inc," 2016. [Online]. Available: www.paradigmoptics.com/. [Accessed: June 6, 2018]; C. Lethien, C. Loyez, J. P. Vilcot, N. Rolland, and P. A. Rolland, "Exploit the bandwidth capacities of the perfluorinated graded index polymer optical fiber for multiservices distribution," *Polymers*, vol. 3, no. 4, pp. 1006–1028, 2011; G. Woyessa, A. Fasano, C. Markos, A. Stefani, H. K. Rasmussen, and O. Bang, "Zeonex microstructured polymer optical fiber: fabrication friendly fibers for high temperature and humidity insensitive Bragg grating sensing," *Opt. Mater. Express*, vol. 7, no. 1, pp. 286–295, 2017.)

through the implementation of a tensile test. By doing that, parameters such as the elastic and plastic limit, Young's modulus, among others, can be accessed. For that reason, the next subsection will address those properties.

1.2.5.1 POFs Mechanical Characterization

One way to find the mechanical properties of a material is through a tensile test, which measures the deformation (δL) of the material at different magnitudes of applied force (F). Depending on the Poisson's ratio, the material can also exhibit a substantial change in its cross-sectional area (A_m). If one assumes that the material does not change its A_m during the tensile test, then, it is possible to define the elongation and force equivalent to:

$$\sigma_e = \frac{F}{A_m}$$

$$\varepsilon_e = \frac{\delta L}{L_0}$$

(1.14)

where σ_e and ε_e define the engineering stress and strain, respectively and L_0 is the initial length of the sample. However, if the material exhibits a significant change in A_m (such as polymer materials), it is preferable to define the true stress (σ_{tr}) and the true strain (ε_{tr}), which are related to the instantaneous dimension of the sample as:

$$\sigma_{tr} = \sigma_e(1 + \varepsilon_e)$$
$$\varepsilon_{tr} = \ln(1 + \varepsilon_e)$$

(1.15)

By plotting the stress as a function of strain, it is possible to find the characteristic stress-strain curve of the material under the experiment, which can then be used to quantify different parameters. Among those is the elastic limit, which is defined as the maximum load that the material can stand and still be able to return back to its original position after the load is removed. Furthermore, in the elastic limit, it can be found the proportional limit that consists on the initial part of the stress-strain curve. The constant of proportionality is called Young's modulus (E) and it is a measure of the stiffness of a material. In some materials, such as glass, the elastic limit coincides with the proportional limit. However, materials like polymers will exhibit two distinct regions governed initially by a linear behavior followed by a nonlinear one. The elastic limit finishes at the so-called yield point, which is defined as the stress and strain value at which the material will deform plastically, meaning that the material will start to deform permanently and non-reversible until breakage. This plastic regime can only be observed in ductile materials like structural steel or polymers, but materials like cast iron or glass show a brittle behavior. Because of that, it is impossible to find the yield point, since the material reaches its rupture limit at the end of the elastic region.

The mechanical properties of optical fibers can be tested through a tensile test machine (one example could be seen in Figure 1.5a, for the Autograph, AGS-5kND, Shimadzu). To hold the fiber terminals to the traction machine clamping tools without the risk of damaging it, aluminum supports containing v-grooves of the same depth of the fiber diameter, can be implemented. The terminals of the fiber samples can then be fixed to the v-grooves by means of an adhesive (see Figure 1.5b), [96].

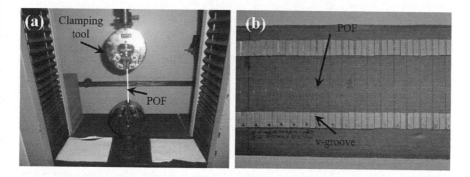

FIGURE 1.5 (a) Clamping tools of the traction machine (Autograph Shimadzu AGS-5kND), used for the stress-strain characterization made in Ref. [96]; (b) POF samples glued at their terminals to aluminum v-grooves.

After the fixation of the aluminum v-grooves in the traction machine clamping tools, the characterization can be initialized. Depending on the fiber drawing conditions (i.e., drawing force, drawing temperature), different degrees of stress frozen in the fiber may be observed. Furthermore, the materials`, as well as the type of microstructure present in the fiber, can also influence the mechanical properties of fibers. As an example, it is shown in Figure 1.6, the typical true stress-strain curves, acquired at constant velocity of 0.6 minute^{-1}, for different PMMA POFs (i.e., SI (MM-MORPOF01 and SM-MORPOF02) and mPOFs (i.e., SM-125, SM-320, MM-150, G3-250)), obtained commercially from different suppliers and with diameters ranging from 115 up to 320 μm.

As can be seen in Figure 1.6, the true stress-strain curves present the typical behavior of a plastic material, where an elastic regime can be observed at the initial stress-strain curve, followed by a plastic regime, until the fiber breaks. However, the curves present different behaviors, Young's modulus, yield point, tensile strength, and failure strain.

In order to determine Young's modulus, the proportional limit of the curves can be confined to the initial part of the curve, in the strain range of 0%–1.3%. For the calculation of the yield point, a conventional strain amount of 0.2% is assumed to produce permanent deformation. Therefore, a parallel line to the one found in the proportional limit is drawn with its origin at 0.2% and slope equal to Young's modulus. The intersection between the two curves is then defined as the yield point (see the example shown in Figure 1.7 for one of the fibers used in the characterization shown in Figure 1.6). The tensile strength and failure strain are calculated as the largest stress and strain, respectively, that the material has experienced during the entire tensile test.

The different mechanical properties observed from different true stress-strain curves can be seen in Table 1.2.

As may be seen from the table and also from Figure 1.6, the failure strain reaches values ranging from ~14% up to 46%. The values are higher than the ones found for uncoated silica fibers which show a brittle behavior with ruptures lower than 1% [97]. Furthermore, values as high as 100% have already been reported for

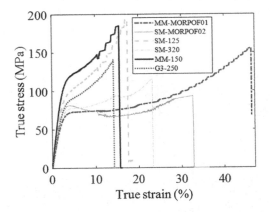

FIGURE 1.6 True stress-strain curves obtained at constant velocity of 0.6 minute^{-1}, for two SI-POFs (MM-MORPOF01 ($2a$ = 250 μm) and SM-MORPOF02 ($2a$ = 115 μm)) and four mPOFs (SM-125 ($2a$ = 125 μm), SM-320 ($2a$ = 320 μm), MM-150 ($2a$ = 150 μm), G3–250 ($2a$ = 250 μm)), with initial lengths of 9 cm.

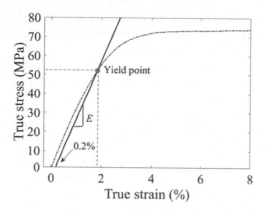

FIGURE 1.7 Method used for the determination of the yield point, executed for the fiber MM-MORPOF01.

TABLE 1.2
Mechanical Properties of Different POFs

	POFs					
Property	MM-MORPOF01	SM-MORPOF02	SM-125	SM-320	MM-150	G3-250
Young's Modulus (GPa)	3.05	3.40	3.87	3.20	4.73	3.50
Yield strength (MPa)	52.15	59.35	70.36	57.30	85.25	63.21
Tensile strength (MPa)	155.29	93.41	194.06	116.06	184.85	140.84
Yield strain (%)	1.83	1.84	1.90	1.94	1.90	1.98
Failure strain (%)	46.25	32.66	17.46	23.22	15.56	14.17

POFs [98]. It can also be seen that the fibers presenting higher failure strains are the ones based on a solid core and cladding (i.e., step-index). The microstructured air holes presented in mPOFs can act as defect centers and probably for that reason they have shown failure strains below the ones found for SI-POFs. The diameter of the fiber, material and microstructure can also influence the stress-strain curve behavior. However, the most important influence can be related to the fiber drawing condition [99]. Literature studies revealed that a fiber drawn under low temperature will need to be pulled with higher drawing force, resulting in an increase of the polymer chain alignment along the fiber length, and thus an increase of the longitudinal mechanical strength [99]. While this allows the creation of a strengthened material, it will also allow the presence of a higher degree of brittleness and high birefringence. One way to decrease the chain alignment is to anneal the fiber below the T_g of the fiber material, allowing a strain relaxation by releasing the uneven stored energy, creating a material more isotropic. Jiang et al. [99, revealed that annealed POFs show lower Young modulus, lower yield strength and tensile strength, but superior ductility.

Annealing POFs also brings additional features as will be explained in Chapter 4. Some of those are linked to the sensitivity in POFBG sensors, in terms of linearity [100–102] and operational temperature [58,103].

Regarding the results shown in Figure 1.6 and the parameters taken from those curves shown in Table 1.2, it can be assumed that the fibers SM-125, MM-150 and G3-250 were drawn under low temperature and high tension. These assumptions are related to the high tensile strength, high yield strength and low ductility. However, the yield strain is similar to all fibers, reaching a value almost close to 2%. Therefore, it is not surprising to find a slightly higher Young's modulus for these three fibers. Young's modulus observed for POFs is much smaller than the ones reported for silica fibers (~70 GPa). Because of that, one may think to use POFs to monitor structures that have itself a low Young's modulus, a condition that the stiff silica fiber could not satisfy since it can act as a local reinforcement structure [104]. Furthermore, when using POFs in strain sensing applications, one needs to keep in mind the value of the yield point, since above this limit, the fiber will deform permanently, affecting also, the light transmission through the fiber [105]. Nevertheless, since POFs can reach high elongations at break [98], they have the potential to be used in large deformations such as the ones found in structural health monitoring [106].

Another key aspect to take into account when designing a POF sensor for stress-strain applications, is the strain rate in which the sensor will be subjected, which is critical in dynamic applications, as we will see later in Sections 4.2 and 5.4. Jiang et al. [99] demonstrated that the yield strength and tensile strength increase with increasing strain rate, achieving a plateau after a specific strain rate. The typical behavior of the stress-strain curve in those conditions may be seen in Figure 1.8, for a POF characterized at strain rates of 0.6, 1.1, and 2.2 minute^{-1}.

The explanation behind the observed behavior is that, at lower strain rates, the polymer material has the natural capability to conform to load, allowing higher ductility and lower strength. On the contrary, with high strain rates, the internal viscosity may not accommodate the rate of loading and the fiber falls in a brittle mode, probably initiated by internal or surface flaws that act as stress concentrators [99].

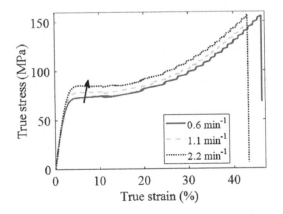

FIGURE 1.8 True stress-strain curves obtained for a POF at strain rates of 0.6, 1.1, and 2.2 minute^{-1}.

1.3 POF END-FACE TERMINATION

The POFs end face quality is one of the main challenges when the splicing process is needed, especially in mPOFs. Thus, the best POF end face must be achieved in order to avoid too much insertion loss and back reflections. For that reason, many authors have been working on this topic, using different techniques, such as the semiconductor dicing (SD) saw [107], the focused-ion-beam (FIB) milling [107] the ultraviolet (UV) laser cleaving [107,108], the hot blade cleaving [109–112], the connectorization process [113,114], the fiber polishing process [109], [115,116] and the liquid nitrogen cleaving method [117]. Among these techniques, the most popular is the hot blade cleaving method, where a dedicated device is used to cut the fiber. The process despite automated, requires a set of properties for proper cleaving a specific fiber type, bringing issues related to the required time to find them. The techniques that use the SD saw, the FIB milling, the UV, and the liquid nitrogen cleaving method, have shown good quality results but only at a laboratory level. Furthermore, the fiber diameter reported on these works is higher than 400 µm, which can be considered large when compared with the values commonly found for SM-POFs (>100 µm). The connectorization process is pointed to be one of the best choices for the POFs end face termination. With this method, the POF can be glued into the bore of a ferrule connector (FC) and then polished, simplifying the coupling process between fibers. However, authors reporting the use of this methodology agree that the fiber core concentricity inside a ferrule connector, is problematic [113]. This relates to the core misalignment, which is due to the pre-etching of the fiber, with regard to fill the ferrule inner hole (bore) [113,114], or can be due to the use of a larger bore ferrule when compared to the POF diameter [118]. In fact, this process presents some problems that can compromise the effectiveness of the technique. This isn't only because of the concentricity of the fiber inside the bore of the ferrule but also due to the concentricity of the core related to the fiber. This last factor plays an important role for POFs since current manufacturing processes do not follow standard procedures, leading to have fluctuations in concentricity and even in dimension of the structures along the fiber length. For MM-POFs, the connectorization process is not a problem because the core is too large. However, when the fiber is SM, the process can be challenging [114]. In addition, the current polishing processes were totally hand-made, which brings imperfections from the user.

1.3.1 CLEAVING PROCESS

The interest in POFs led to the development of tools for their cleavage. These tools are in its simplest form, based on a razor blade that is pushed against the transversal section of a POF. As an example of those devices, it can be found the POF cutter block and the hot knife, as shown in the examples displayed in Figure 1.9a and b. Here the hot knife also consists of a heating system fixed to the blade, in order to soften up the POF while it is being cleaved.

While the devices presented in Figure 1.9 can effectively cleave POFs of diameters up to 1,000 µm, they produce POF end face terminals with poor quality, associated with the lack of control on the temperature of the blade and POF, as well as lack

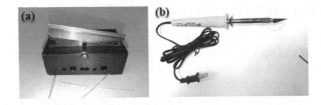

FIGURE 1.9 Commercial available tools for cutting POFs: (a) POF cutter block and
(b) hot knife.

on the control on the velocity in which the blade passes through the fiber. For larger
core fibers, the quality of the POF end face is not critical, since the defects created
on the surface are small when compared with the diameter of the POF. However,
when dealing with POFs in which the core can be lower than 50 μm, sometimes
even lower than 8 μm in the case of SM-POFs, the fiber termination needs to be
of high quality, necessary to have low insertion loss on the fiber coupling process.

1.3.2   ELECTRONICALLY CONTROLLED POLYMER OPTICAL FIBER CLEAVER

POFs can exhibit different mechanical properties, depending on the drawing temperature, drawing speed, fiber structure, material composition, etc. Therefore, the
blade penetration on the POF will depend on these parameters and different cleaving scenarios may occur, depending if the fiber is brittle or ductile [110]. In the
case of brittle POF, a high density of stress at the edge of the blade will occur,
leading to the formation of a crack. However, in a POF with ductile characteristics,
the density of stress at the edge of the blade is not sufficient to allow the crack
formation and thus the blade will always be in contact with the polymer material.
However, if temperature is used, the phase transitions from brittle to ductile can be
achieved. For that, one needs to adjust the temperature regarding the POF material,
structure and thermal history (drawing temperature and post annealing process).
Furthermore, the velocity in which the blade is pushed against the fiber should also
be taken into account. In this way, an optimization process needs to be done for
each specific fiber. Because of that, different authors have already demonstrated
the control of those parameters to achieve high-quality POF end face terminals
[109–111]. An example of an automated POF cleaving machine can be seen in
Figure 1.10.

The automated cleaving systems are based on a motorized linear stage used to
move the blade with a specific velocity against the POF. The blade is secured in a
heating element that is fixed to the motorized linear stage. Its sharpness is essential
and blades with long flat-edge (conventional razor blades are purely wedge-shaped)
offer best results. Identically, at the bottom, there is a similar heating system, where
the blade is now replaced by a v-groove plate that is used to secure the fiber and to
give a 90° angle between the longitudinal axis of the POF and the edge blade. The
linear stage in the bottom part is used to move the base to another position, allowing
to avoid blade damages and the polymer debris left from the last cleavage, which can
have a negative influence on the subsequent cleavage.

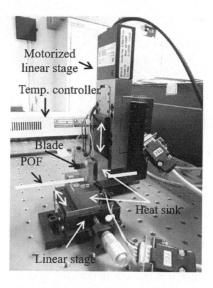

FIGURE 1.10 Electronically controlled polymer optical fiber cleaver, constructed based on the works developed in Refs. [110,111].

For each type of polymer optical fiber, an empirical process needs to be performed in order to access the right temperature of the blade and plate where the fiber is secured, as well as the velocity in which the blade is pushed against the POF. During the process, different types of imperfections could be observed, such as burn of the fiber end face, core shift, crack formation rough end face, etc. Some of those imperfections may be seen in Figures 1.11a–d.

Since the process requires an empirical trial and error test before reaching the proper cleaving parameters, it may be considered a time-consuming process. Nevertheless, after choosing the right parameters, POFs of the same spool may be readily cleaved with good quality. Examples of fibers cleaved with proper parameters may be seen in Figure 1.12.

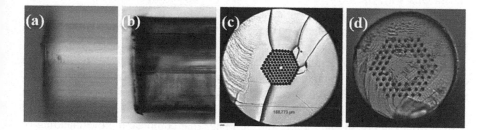

FIGURE 1.11 Transversal ((a), (b)), and top view ((c), (d)), microscope images of mPOFs, presenting different types of defects due to improper cleaving parameters (i.e., temperature of the blade and plate in which the fiber is secured, as well as velocity in which the fiber is pushed against the POF). (a) POF terminal with surface burned, (b) core shift, (c) crack formation and (d) rough surface.

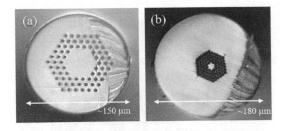

FIGURE 1.12 Top view of mPOFs cleaved with an automated cleaving method using proper parameters. PMMA mPOFs: MM (a); few-mode (b); prepared using a blade velocity of 2 mm/second, blade temperature of 65°C and plate temperature of 75°C.

Another interesting method to cleave POFs has been also explored by Sáez-Rodriguez et al. [119,120], in which a time-temperature equivalent principle is used to cleave a POF at room temperature by applying a sawing motion instead of chopping as performed by the hot blade cleaving methods. However, as it occurs with the methodology described before, an optimization procedure (i.e., the velocity of the blade against the POF and blade angle) also needs to be performed for each POF sample.

1.3.3 POLYMER OPTICAL FIBER CONNECTORIZATION

One interesting POF end face termination process commonly used by many authors is the fiber connectorization process [113,114,121]. This procedure has been adopted from the already developed connectorization procedure developed for silica optical fibers. In this process, the fiber is inserted in the bore of a ferrule connector that is pre-filled with a resin. After the resin is set, the fiber tip is cleaved and the polishing procedure follows. For that, different polishing films are used in descending order of grain sizes. The connectorization process is already standardized for silica fibers, where bore ferrule diameters of ~125 μm are used to hold the standard silica optical fiber (diameter of ~125 μm), with the core in a concentric position. However, POF technology is still immature and POFs may be found in a variety of diameters. Nevertheless, the diameter fluctuation along the fiber length is also present. In this way, the POF preparation through the connectorization process faces some challenges namely for the matching between the fiber diameter and the commercially available bore diameter ferrules. To suppress this problem an etching of the POF could be performed, in order to reach the diameter of the bore ferrule [113,114].

Depending on the fiber material, different solvents may be used to reduce the diameter of a POF. For instance, to reduce the diameter of PMMA-based POF, acetone [94,113], [122], mixtures of acetone in water [94] or also mixtures of acetone and methanol are normally used [94,123]. TOPAS®-based fibers have been etched with cyclohexane [124] and the PC and polyester that compose the overcladding of the commercial GI-POF from Chromis Fiberoptics Inc., with chloroform [125].

In order to estimate the etching time needed for a given POF to reach a specific diameter, the etching rate needs to be accessed. One example of the diameter evolution versus exposure time, for a PMMA mPOF with initial diameter of ~155 μm, and immersed in acetone, may be seen in Figure 1.13.

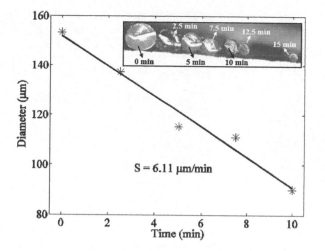

FIGURE 1.13   Etching rate of a PMMA mPOF immersed in a solution of acetone at room temperature. The inset figure shows the pictures of the fibers for each corresponding time.

The etching rate achieved for the POF presented in Figure 1.13 is about 6.1 µm/minute, and thus to achieve a case example of a bore ferrule diameter of ~126 µm, it is necessary to immerse the fiber for about 4 minutes. Since the tolerance between the fiber and bore ferrule needs to be tight in order to achieve good concentricity results (mostly important for SM fibers, that present small core diameters), the etching process needs to be performed with accuracy and precision.

After the etching process, the fiber is inserted into the bore ferrule that has been previously filled with resin. The resin will cure and the fiber tip can be cleaved with a hot blade close to the ferrule tip. The temperature will avoid the formation of cracks that can propagate through the longitudinal axis of the fiber to the inner region of the bore ferrule. Furthermore, the cleavage needs to be performed a few tens of microns far from the ferrule tip to avoid the formation of cracks or core shift too close to the ferrule tip.

The first report of a connectorized POF was hand-made and performed by fixing the ferrule connector in a polishing disc and start polishing the fiber tip in a figure-eight pattern, for different grain polishing films [113,114]. However, commercially available polishing machines can be used to perform the process in a more automated fashion, offering low operator skill, low processing time and high-quality termination. An example of such a device may be seen in Figure 1.14a (REV™ Connector Polisher, from Krell Technologies, Inc.), while the prepared connectors containing POFs may be seen in Figure 1.14b.

The polishing procedure can follow the same steps as the ones mentioned for silica fibers; however, due to the softness of POFs, only two cycles of 15 seconds are needed. The first cycle is made with a large grain size polishing film (>3 µm), which is used to roughly scratch the fiber, removing the imperfections and flattening the fiber to the tip of the ferrule. The second cycle is made with a small grain size polishing film (e.g., <1 µm), which allows smoothing the fiber end face. Water spray can

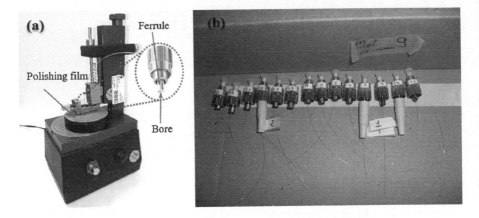

FIGURE 1.14   (a) Connector polishing machine. (b) Several POFs glued to different ferrule conectors with physical contact (FC/PC) and prepared with the polishing machine.

be used in this final process to remove fiber and resin debris left from the polishing procedure. An example of the final results produced through this procedure may be seen in Figure 1.15a–e for mPOFs and (f) for a SI SM-POF.

Despite the absence of roughness, scratches, cracks and core-shift in the microscopic images shown in Figure 1.15, all the fibers share core concentricity issues. In the case of the microscopic images shown in Figures 1.15a, c, and d, the POFs have been etched prior to the connectorization process, in order to fit the diameter of the bore ferrule and as can be seen, there exists a large gap between the fiber and the

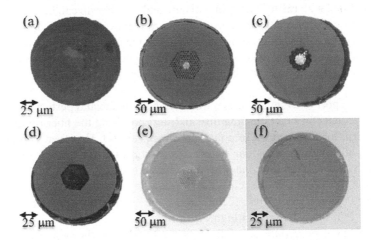

FIGURE 1.15   Microscope images of different POF terminals in FC/PC connectors, prepared through the polishing procedure. The fibers shown in (a)–(e) are mPOFs, while the one shown in (f) is SI. The bore ferrule diameters are 250 μm for (b) and (e), while 125 μm for the rest of the prepared fibers. Fibers presented in (a), (c) and (d) were slightly etched prior to insertion of the POF into the bore ferrule.

bore ferrule, associated to the inherent difficulty in finding the right POF diameter during the etching procedure. Nevertheless, in POFs with large core diameter (typically with MM behavior), the core concentricity is not problematic in the coupling process [118], as it is indicated for the fibers shown in Figure 1.15a–c, which have the capability to still guide light after the coupling process (considering that the other POF terminal was well terminated and butt-coupled to a silica SM fiber). However, for thin diameter core fibers (i.e., fibers with few-mode (FM) and SM behavior), as it is the case of POFs shown in Figure 1.15d–f, the concentricity is highly important for the efficiency of the light coupling. Thus, in those cases, the light coupling using fiber optic connectors is not allowed, unless tight tolerances are met during the etching procedure.

1.3.4 FIBER POLISHING PROCESS

Considering the previous sections, the proposed fiber polishing process is the combination of both hot cleaving and connector polishing processes. However, the techniques are manipulated in order to solve the drawbacks associated with the time dependence and core misalignment for cleaving and connector polishing procedures, respectively. The technology has been proposed by Oliveira et al. [115], and it provides high-quality POFs end face; low operator skill; low processing time, repeatability, among others. The initial step required for this process is to find bore ferrules with diameters that closely match the ones of the fiber to be prepared. Next, the POF is inserted in the bore ferrule and cleaved by hand with a hot blade in which the temperature is controlled to be within 70–80°C, high enough to prevent crack formation but at the same time below the melting temperature of the polymer fiber materials. The ferrule is inserted in the polishing machine and the fiber is pushed down against the polishing film during the entire polishing process. Since the fiber can move freely inside the bore ferrule, it is important to have similar diameters of the bore ferrule and fiber. During the fiber end face preparation process, it is important to check regularly the quality of the fiber tip in order to verify the presence of defects. This can be easily accessed through the use of an optical fiber scope (e.g., OFS300-200C) allowing to take the necessary actions before the proceeding polishing step. The work performed by Oliveira et al. was done using fibers of different diameters, ranging from 125 to 500 µm, made of different materials, structures and suppliers. Unprecedented results regarding the microscopic images of the prepared end face POF terminals can be seen in Figure 1.16. As can be seen, the fibers terminals appear without defects associated to the blade, since the first polishing procedure removes a great amount of the POF terminal and inherently the large scale defects (mainly the core shift and longitudinal crack formation). Furthermore, defects associated with scratches are easily removed by the second polishing procedure that is used to smooth the fiber end face. The process is fast, reliable and semi-automatic avoiding human errors and offering unprecedented end face POF results [115]. The prepared POFs are well suited for applications regarding free space coupling to other POFs or silica fibers, solving problems related with the concentricity of the core in POFs and POF connectorization procedures.

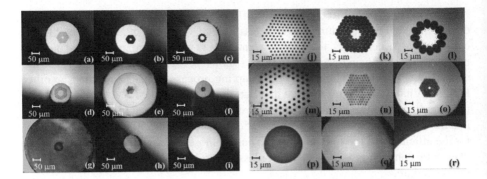

FIGURE 1.16 Microscope images of different POFs, acquired with magnification of 50×
(a)–(i) and 200× (j)–(r), for: mPOFs (a)–(f), (j)–(o); GI-POFs (g),(p); and SI-POFs (h), (i) and
(q), (r). The fibers are from different suppliers and have diameters ranging between 125 and
500 μm. (Due to the dimensions of the fibers, the microscope images shown in (p) and (r)
only show the core and core-cladding interface, respectively.) (Adapted from R. Oliveira,
et al., "Smooth end face termination of microstructured, graded-index, and step-index polymer optical fibers," *Appl. Opt.*, vol. 54, no. 18, PP. 5629–5633, 2015, with permission from the
Optical Society of America.)

1.4 SILICA TO POLYMER OPTICAL FIBER CONNECTORIZATION

Currently, most of the optical equipments, such as sources and detectors, are terminated with silica optical fibers. Therefore, when working with POFs, it is sometimes
necessary to create a splice between these two fibers. In fact, this has been the main
obstacle in the development of the POF technology. In fact, the splice needs to ensure
the best mechanical stability and the lowest possible loss. The latter is the key aspect
since polymer materials can impose a great amount of attenuation even working at
the low loss region of polymers.

The most known technique used to produce a permanent and stable connection
between fibers is fusion splicing. This involves heating the fiber ends that were previously cleaved, in order to fuse them. The process can be done through an electric arc
discharge or through CO_2 laser irradiation. Considering silica to POF splice, these
methods cannot be applied since the glass transition temperature of silica is ten times
higher than that of polymers.

Another widely spread technique used to make a permanent connection between
two silica fibers is based on mechanical splices. In general, higher losses and greater
reflectance than fusion splices may be observed. The technique relies on the use of
a precision alignment mechanism where the fibers are inserted with a drop of index
matching oil or a transparent optical resin in-between. The effectiveness of the alignment is always related to the core concentricity and cladding diameter of the fibers. If
one of the cores is not concentric, the light coupling can be compromised. The same
applies to the cladding diameter that needs to be the same for both fibers and also
for the alignment device. For POFs, this cannot be applied by the reasons explained
before such as poor core concentricity and large cladding diameter variations along
the fiber length, making the technique infeasible.

Fiber optic connectors are a possibility for semi-permanent connections. These devices are largely used for an easy connection that needs to be released more often. There are a large number of connector types and the preparation of such connectors was explained in the previous section. The concentricity of the fiber in the bore ferrule is an issue for fibers with small cores, compromising the coupling efficiency between fibers. The technique has been proposed in different works [113,114,118,126,127], most of them using POFs with multimode behavior because of the lower tolerances on the coupling efficiency associated with the large core diameter. Yet, it is still possible to find some publications reporting the use of POFs with SM behavior prepared through this methodology [114,126,127].

Another way to couple light from one fiber to another is through free space optics [118,128,129]. For that, a system of lens may be used, having also the possibility to insert other optical elements, such as filters and isolators. An advantage of this methodology is related to the high coupling efficiency that can be achieved, even for fibers with different mode field diameters. However, the system needs to be maintained in a stable configuration and away from dust. Additionally, the cost for a single connection and the need to maintain all the parts in laboratory makes this coupling process impractical for in-field applications. Nevertheless, there are works reporting this process to couple light from silica fibers into POFs [130].

Tapered and lensed fibers are good solutions for coupling light between optical fibers. Indeed, the use of a tapered silica fiber to couple light into a microstructured POF has been recently proposed [131]. The fibers were mechanically aligned in a fusion splicer, and the tip of a tapered silica fiber was inserted in one of the holes of the mPOF, close to the core fiber, allowing mechanical stability. Despite the promising results, the connection losses were relatively high, but still acceptable for sensing solutions [131].

Finally, the most preferred method to couple light from a silica fiber to POF is the butt coupling process. In this technique, the fibers are axially aligned through their cores and an index matching gel with a refractive index similar to that of the fibers, is used between the fiber terminals. Normally, when it is necessary a stable connection with this technique, the index matching gel is replaced by an optical adhesive that is hardened after UV exposure. Additionally, to allow better mechanical stability, the splice is sometimes protected with additional layers of glue [132,133], or even through the use of capillary tubes filled with resin [94].

1.4.1 COUPLING LOSSES

Splicing optical fibers always involve dealing with loss. Consequently, it is necessary to know which parameters affect the splicing loss in order to keep it at a minimum level. Generally, the splicing defects can be summarized as longitudinal fiber separation, tilt between the fibers, transversal offset and spot size mismatch [134,135]. The different splicing defects can be seen in Figure 1.17.

For a SI fiber with a Gaussian profile, the mode field radius (w) can be calculated through Ref. [134], as:

$$\frac{w}{a} = 0.65 + \frac{1.619}{V^{3/2}} + \frac{2.879}{V^6} \tag{1.16}$$

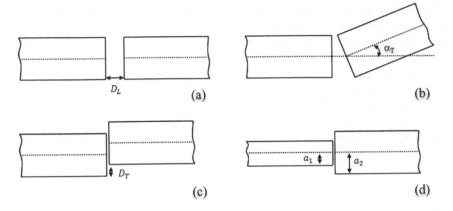

FIGURE 1.17 General splicing defects. Fibers separated in longitudinal direction (a), tilted (b), or with offset (c) with respect to each other. Fibers with different core radius (d).

where for an mPOF, w is expressed through Ref. [136], as:

$$\frac{w}{a} = \frac{A}{V_{mPOF}^{2/(2+g)}} + \frac{B}{V_{mPOF}^{3/2}} + \frac{C}{V_{mPOF}^{6}} \tag{1.17}$$

where A, B and C are the fitting parameters which are 0.7078, 0.2997, and 0.0037, respectively, while V_{mPOF}, refers to the normalized frequency found for microstructured fibers, normally refereed as photonic crystal fibers (PCFs), defined as [137]:

$$V_{mPOF} = \frac{2\pi\Lambda_d}{\lambda}\sqrt{n_{co}^2 - n_{cl}^2} \tag{1.18}$$

where Λ_d is the hole to hole distance and n_{co} and n_{cl} are the effective refractive index of the core mode and of the fundamental space-filling mode in the triangular air-hole lattice, respectively.

Following Equations 1.16 and 1.17, it is possible to estimate different splice losses. When an optimum Gaussian field distribution with radius w_1 and core radius a_1 enters into the input plan of the second fiber with an optimum Gaussian field distribution with radius w_2 and core radius a_2, a mode matching will occur. The calculations for such a Gaussian field mode matching are found in Ref. [135] and the following equations will only show the final result for each particular fiber coupling misalignment. Considering the case where there is longitudinal displacement D_L (see Figure 1.17a), one can find the power transmission coefficient as [134]:

$$T = \frac{4\left(4Z^2 + \dfrac{w_1^2}{w_2^2}\right)}{\left(4Z^2 + \dfrac{w_1^2 + w_2^2}{w_2^2}\right)^2 + 4Z^2\dfrac{w_2^2}{w_1^2}} \tag{1.19}$$

where Z defines the normalized fiber separation, expressed as:

$$Z = \frac{D_L}{n_{ext} k w_1 w_2}$$

(1.20)

where n_{ext} is the index of the medium where the fibers are immersed and k the wavenumber. Regarding fibers tilted in relation to each other (see Figure 1.17b), with an angle α_T, the power transmission coefficient may be expressed as:

$$T = \left(\frac{2w_1 w_2}{w_1^2 + w_2^2} \right)^2 \exp\left(-\frac{2(\pi n_{ext} w_1 w_2 \alpha_T)^2}{(w_1^2 + w_2^2)\lambda^2} \right)$$

(1.21)

For a transverse displacement (D_T) between the fibers, as shown in Figure 1.17c, the power transmission coefficient is described as:

$$T = \left(\frac{2w_1 w_2}{w_1^2 + w_2^2} \right)^2 \exp\left(-\frac{2D_T^2}{w_1^2 + w_2^2} \right)$$

(1.22)

If it is considered a perfect alignment splice, D_L, α_T, and D_T are equal to zero and thus Equations 1.19, 1.21, and 1.22, will be expressed as:

$$T = \left(\frac{2w_1 w_2}{w_1^2 + w_2^2} \right)^2$$

(1.23)

This specific case can be seen in Figure 1.17d, where the fibers are considered to have different core radius. The relation between the transmission coefficient and the ratio of the mode field radius (w_1/w_2) can be seen in Figure 1.18.

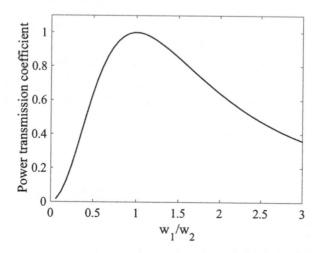

FIGURE 1.18 Power transmission coefficient as function of w_1/w_2 for a perfect alignment splice.

As can be seen from Figure 1.18, the maximum transmission is achieved when the Gaussian mode field radii are similar. It is important to note that the function is same if T is expressed as a function of w_2/w_1. In order to have a practical example, let's consider the coupling loss at the 1,550 nm region from a standard silica SMF-28e to different POFs sold from different suppliers. The fiber properties, together with its theoretical coupling loss to a SMF-28e can be seen in Table 1.3.

The theoretical coupling loss shown in Table 1.3 reveals that higher losses are observed for the larger difference between the mode field radius, as expected. From the different fibers under analysis, the one having higher coupling loss is the mPOF, where the mode mismatch is significant. However, the losses reduce to zero when there is a perfect mode matching, as is the case presented for the coupling between fibers of the same type (i.e., SMF-28e to SMF-28e).

When the splice misalignments are taken into account, the model becomes more complex. Thus, in order to observe the evolution of the power transmission coefficient with each particular splice misalignment, let's consider the simplest case, where the mode field radius is the same ($w_1 = w_2 = w$). Then, Equations 1.19, 1.21 and 1.22 can be rewritten respectively as:

$$T = \frac{1}{1+Z^2} \tag{1.24}$$

$$T = \exp\left(-\left(\frac{D_T}{w}\right)^2\right) \tag{1.25}$$

$$T = \exp\left(-\left(\frac{\pi n_{ext} w \alpha_T}{\lambda}\right)^2\right) \tag{1.26}$$

To have a better visualization of the influence of the fiber misalignments on the transmission coefficient, it was considered the coupling between two SI-POFs based

---

TABLE 1.3

Fiber Properties and Coupling Loss to a SMF-28e at 1,550 nm Region

Fiber	Manufacture	Type	Material	a (μm)	NA	λ_c(nm)	V	w (μm)	Coupling Loss (dB)
SMF-28e	Corning®	SI	Silica	4.1	0.12	1,305	1.99[a]	5.21[b]	0.00
SM-MORPOF03	Paradigm Optics, Inc.	SI	PMMA	4.0	0.07	750	1.14[a]	13.34[b]	3.38
SM-MORPOF02	Paradigm Optics, Inc.	SI	PMMA	1.6	0.27	1,100	1.72[a]	2.32[b]	2.57
SM-125	Kiriama, Ltd	mPOF	PMMA	2.0	0.22	None	2.60[c]	1.31[d]	6.49

[a] Calculated using Equation 1.1.
[b] Calculated using Equation 1.16.
[c] Calculated using Equation 1.18, considering $\Lambda_d = 2.9$ μm.
[d] Calculated using Equation 1.17, considering $g = 8$, [136].

on the fiber SM-MORPOF03, described in Table 1.3. For that, it was used one of the regions where PMMA shows lower losses (λ = 850 nm), producing V = 2.07 and w = 4.92. Results concerning the evolution of T with fiber displacement, lateral offset, and tilt can be seen in Figures 1.19–1.21 respectively.

From Figure 1.19, it can be observed two solutions obtained when n_{ext} = 1 for the fiber coupling in air and n_{ext} = 1.485, when the fibers tips are immersed in an index matching oil with refractive index similar to that of the POFs at the 850 nm region. For both cases, it is evident that the fiber displacement has low influence on T.

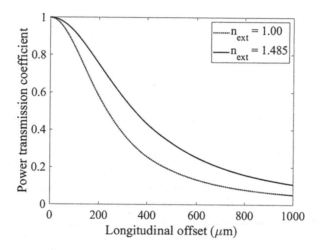

FIGURE 1.19 Power transmission coefficient as a function of the longitudinal offset, considering the coupling between two SI-POFs (SM-MORPOF03), in air and immersed in index matching oil.

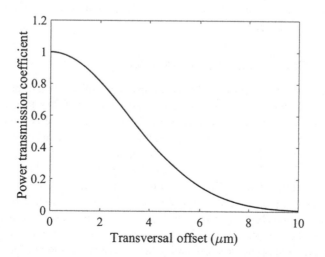

FIGURE 1.20 Power transmission coefficient as a function of the transversal offset between two SI-POFs (SM-MORPOF03).

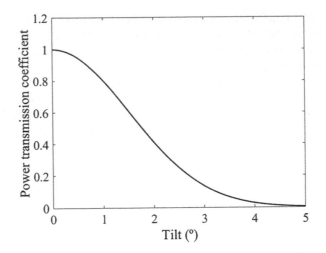

FIGURE 1.21 Power transmission coefficient as a function of the tilt between two SI-POFs (SM-MORPOF03).

Additionally, it can be observed that when the fiber coupling is in air, the transmission coefficient is reduced compared to that immersed in an index matching oil.

Figure 1.20 shows that T can drop drastically as the fiber transversal offset increases, reaching half of its transmission for a lateral misalignment of ~4 µm, which corresponds to the radius of the fiber.

Regarding Figure 1.21, it can be seen that the transmission between the fibers has high influence with the angle between the fibers, where an angle of only 1.77°, reduces the power transmission to half of its original value.

The discussion presented here was focused on a particular coupling misalignment. For the general case, where more than one coupling misalignment is present, the equations shown above will become much more complex and a detailed explanation of that may be found in Ref. [138].

1.4.2 BUTT COUPLING/UV CURING PROCESS

Due to the absence of a commercial device capable to splice silica to polymer optical fibers and taking into account that the connectorization of POFs in ferrule connectors produces cores that are off-center in the bore ferrule, there is a requirement of a new methodology capable to produce a stable, robust and reliable splicing process between these two dissimilar fibers. This can be accomplished through the use of a "cold splice" [139], in which the fibers are axially aligned through their cores and then, mechanically stabilized with UV resin. For that, the silica optical fiber is normally cleaved with an 8° angle in order to prevent Fresnel reflections, while the POF can be prepared through the cleaving or polishing process described in the previous section. The fibers can then be placed in an alignment system, capable to assist the coupling process. An example of the experimental arrangement is depicted in Figure 1.22.

The alignment of the fibers can be performed through the use of a 3D mechanical positioner. Furthermore, it can be made in real-time with the help of two cameras

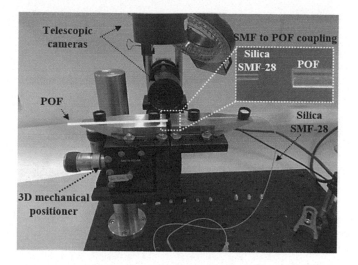

FIGURE 1.22 Setup used to provide axial and longitudinal alignment between silica and polymer optical fibers. The inset shows a silica optical fiber at the left, being aligned to a large diameter POF at the right.

positioned orthogonally to the fibers. In most splicing systems, precise fiber alignment is usually done through the analysis of the position of the core and cladding of the fibers. However, when dealing with microstructured fibers, the overlap of the holes over the core area makes it difficult to identify the core through digital images. Additionally, the refractive index contrast in some POFs makes it difficult to find the core. Nevertheless, in most of the cases, the concentricity of the core POF varies along the length of the fiber. For that reason, the alignment of the fibers can be assisted at the POF far end, either by measuring the power transmitted through the fiber [116] or by checking the near field pattern of the POF [115]. After aligning the fibers in the three orthogonal axes, a drop of an index matching gel or UV resin (in case of a permanent splice), with refractive index similar to one of the fibers involved in the coupling, is carefully inserted between the fibers. The distance in-between needs to be minimized in order to avoid the formation of Fabry-Pérot (FP) cavities that can deteriorate the transmission signal. The complete splice process captured at different time instants for one of the side cameras shown in Figure 1.22 can be visualized in Figure 1.23.

The confirmation of the core alignment may be done through the visualization of the near field image, which can be accomplished by placing an objective lens together with a laser beam profiler at the far end of the POF. An example of near field images taken for different fibers, before and after the core alignment, may be seen in Figure 1.24a–i and j–r, respectively.

In case of a permanent mechanical splice, most desired for field applications, the resin should be carefully chosen. In literature, resins capable to polymerize when exposed to UV radiation are the most preferred, being, Norland® [132,140], Henkel® [133,141] and DYMAX® [139,142,143], the suppliers mostly reported on literature. Nevertheless, adhesives that do not require light exposure for the curing process

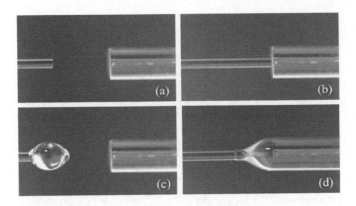

FIGURE 1.23 Steps needed for splicing a silica optical fiber (left) to a POF (right). (a) Fiber approximation, (b) transversal alignment, (c) insertion of photopolymerizable resin, (d) longitudinal alignment and UV curing.

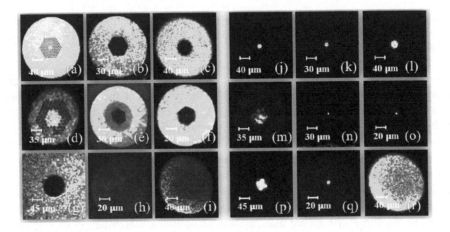

FIGURE 1.24 POFs near field images, without (a)–(i) and with (j)–(r) correct core alignment to a pigtail silica SM fiber. The images shown in (a)–(f), (j)–(o) are referred to mPOFs; (g), (p) to GI-POF; and (h), (i) and (q), (r) to SI-POFs. (Adapted from R. Oliveira, et al., "Smooth end face termination of microstructured, graded-index, and step-index polymer optical fibers," *Appl. Opt.*, vol. 54, no. 18, PP. 5629–5633, 2015, with permission from the Optical Society of America.)

have also been reported for the production of permanent splices [139]. However, they require longer curing times, which makes their use infeasible.

Among the different characteristics that a photopolymerizable resin should have, the most important ones are:

- the mechanical resistance, allowing to have a secure connection between the fibers;
- the transparency, providing light guidance with minimum insertion loss; and
- the curing time, which needs to be fast enough for mass production.

Other parameters, such as refractive index, viscosity, operational temperature, moisture resistance, etc., need also to be taken into account. In order to make a comparison, the properties of different UV photopolymerizable resins commonly used for the silica to polymer optical fiber splicing process, namely: NOA76 [144,145]; NOA78 [132,146]; NOA86H [147,148]; Loctite3525 [133,143,149,150], are shown in Table 1.4.

From Table 1.4, it is possible to see slight variations between the refractive index of the cured resins. Indeed, the resin selection needs to have a cured refractive index that closely matches the one of the fibers used (either silica or POF). This can be concluded from Figure 1.19 where it can be seen an enhancement on the power transmission coefficient when the refractive index closely matches that of the POF. This will allow a smooth light transition from one medium to another, avoiding the formation of FP cavities as well as reducing back reflections at the POF end face terminal and consequently the insertion loss [139]. Another property to take into consideration when comparing different resins is the mechanical strength that they can offer. Thus, from Table 1.4, it is obvious that the UV resin that provides higher mechanical resistance is NOA 86H. Because of that, this UV resin has been reported to splice silica and polymer fibers and also to be used for strain sensing applications [148,153,154]. Despite the lower resistance presented by other UV resins, different authors bypass the problem either by depositing several layers of UV curable resin, surrounding the fiber coupling [133]; through the use of a second UV curable resin with a higher strength [146]; and by depositing an acrylic sealant with humidity resistance [132]. Even so, the splice is always a critical part of the fiber-based device and thus in order to allow mechanical stability and to provide robust protection of the silica to POF splice, there are reports on the use of capillary tubes filled with epoxy resin at the splice region [94,123]. An example of this type of fiber coupling can be seen in Figure 1.25.

Another relevant parameter presented in Table 1.4 is the viscosity of the glue. This is particularly important when working with mPOFs, where the glue can infiltrate

TABLE 1.4
UV Photopolymerizable Resins Characteristics

Company & Resin Reference	Refractive Index @ 589 nm (cured)	Tensile Strength (MPa)	Viscosity @ 25°C (cps)	Temperature Range (°C)	Recommended Use
Norland® NOA 76 [151]	1.51	3.1	3,500–5,500	−15 to 60	Ideal for bonding glass to plastic
Norland® NOA 78 [151]	1.50	4.5	8,000–11,000	−20 to 60	Ideal to bond plastic to glass
Norland® NOA 86H [151]	1.55	54.0	200–300	−125 to 125	Good adhesion, solvent resistance
Henkel® LoctiteAA-3525 [152]	1.51	24.1	15,000	−54 to 149	High curing speed; durability to moisture

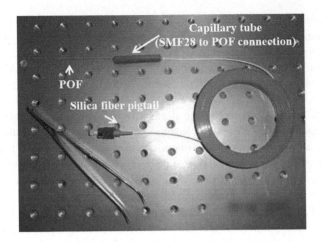

FIGURE 1.25 POF spliced to a silica pigtail fiber. The splice is made of UV photopolymerizable resin (NOA78), which has low tensile strength, and because of that, it is reinforced with a capillary tube that is filled with high strength epoxy resin.

the holes if the viscosity is too low [133]. As a consequence, the guiding properties of the fiber where the glue infiltrates is compromised, leading to higher losses at the splice. Therefore, the resins with a higher viscosity, such as the NOA 78 and the LoctiteAA-3525, are expected to solve the glue infiltration through the holes, since they are less prone to the capillary action. While this could be problematic for fibers prepared by the cleaving method because the holes are completely opened in these fibers, the same does not apply for fibers prepared through the polishing method. In this methodology, the holes are clogged during the polishing procedure with remains of polymer debris [154]. For POF structures other than mPOFs, the viscosity is not a concern and any of the resins shown in Table 1.4 can be used.

For the operational temperature, one can find that the glues, such as NOA 86H and the LoctiteAA-3525, offer a higher temperature, and thus, they are the preferred choice for high-temperature applications. Other features that could benefit different applications and that are present in some special UV curable resins are the durability to moisture and the chemical resistance. Thus, they should be taken into account when the device is intended to work in those applications. One example was presented by Oliveira et al. who reports the use of fiber splices made of different photopolymerizable resins for the detection of RH [155,156].

REFERENCES

1. J. Hecht, *City of Light: The Story of Fiber Optics*. New York: Oxford University Press, Inc., 1999.
2. Dupont, "Low Attenuation All-Plastic Optical Fibre," GB2006790B, 1979.
3. Dupont, "Low Attenuation Optical Fibre of Deuterated Polymer," GB2007870B, 1978.
4. Mitsubishi Rayon Co. Ltd., "Light Transmitting Fibres and Method for Making Same," GB1431157A, 1974.
5. Mitsubishi Rayon Co. Ltd., "Light Transmitting Filament," GB1449950A, 1974.

6. A. W. Snyder and J. D. Love, *Optical Waveguide Theory*, 2nd ed. Boston, MA: Springer, 1984.

7. M. G. Kuzyk, U. C. Paek, and C. W. Dirk, "Guest-host polymer fibers for nonlinear optics," *Appl. Phys. Lett.*, vol. 59, no. 8, pp. 902–904, 1991.

8. D. Bosc and C. Toinen, "Full polymer single-mode optical fiber," *IEEE Photonics Technol. Lett.*, vol. 4, no. 7, pp. 749–750, 1992.

9. G. Woyessa, A. Fasano, A. Stefani, C. Markos, H. K. Rasmussen, and O. Bang, "Single mode step-index polymer optical fiber for humidity insensitive high temperature fiber Bragg grating sensors," *Opt. Express*, vol. 24, no. 2, pp. 1253–1260, 2016.

10. G. Zhou, C. F. J. Pun, H. Y. Tam, A. C. L. Wong, C. Lu, and P. K. A. Wai, "Single-mode perfluorinated polymer optical fibers with refractive index of 1.34 for biomedical applications," *IEEE Photonics Technol. Lett.*, vol. 22, no. 2, pp. 106–108, 2010.

11. G. D. Peng, P. L. Chu, Z. Xiong, T. W. Whitbread, and R. P. Chaplin, "Dye-doped step-index polymer optical fiber for broadband optical amplification," *J. Light. Technol.*, vol. 14, no. 10, pp. 2215–2223, 1996.

12. J. Bonefacino et al., "Ultra-fast polymer optical fibre Bragg grating inscription for medical devices," *Light Sci. Appl.*, vol. 7, no. 3, p. 17161, 2018.

13. H. Y. Tam, C. F. J. Pun, G. Zhou, X. Cheng, and M. L. V. Tse, "Special structured polymer fibers for sensing applications," *Opt. Fiber Technol.*, vol. 16, no. 6, pp. 357–366, 2010.

14. J. Yu, X. Tao, and H. Tam, "Trans-4-stilbenemethanol-doped photosensitive polymer fibers and gratings," *Opt. Lett.*, vol. 29, no. 2, pp. 156–158, 2004.

15. J. M. Yu, X. M. Tao, and H. Y. Tam, "Fabrication of UV sensitive single-mode polymeric optical fiber," *Opt. Mater.*, vol. 28, no. 3, pp. 181–188, 2006.

16. W. Wu et al., "Design and fabrication of single mode polymer optical fiber gratings," *J. Optoelectron. Adv. Mater.*, vol. 12, no. 8, pp. 1652–1659, 2010.

17. G. D. Peng, Z. Xiong, and P. L. Chu, "Photosensitivity and gratings in dye-doped polymer optical fibers," *Opt. Fiber Technol.*, vol. 5, no. 2, pp. 242–251, 1999.

18. Y. Luo, Q. Zhang, H. Liu, and G.-D. Peng, "Gratings fabrication in benzildimethylketal doped photosensitive polymer optical fibers using 355 nm nanosecond pulsed laser," *Opt. Lett.*, vol. 35, no. 5, pp. 751–753, 2010.

19. Z. Li, H. Y. Tam, L. Xu, and Q. Zhang, "Fabrication of long-period gratings in poly(methyl methacrylate-co-methyl vinyl ketone-co-benzyl methacrylate)-core polymer optical fiber by use of a mercury lamp," *Opt. Lett.*, vol. 30, no. 10, pp. 1117–1119, 2005.

20. T. Wang et al., "Enhancing photosensitivity in near UV/vis band by doping 9-vinylanthracene in polymer optical fiber," *Opt. Commun.*, vol. 307, pp. 5–8, 2013.

21. J. Bonefacino, X. Cheng, M.-L. V. Tse, and H.-Y. Tam, "Recent progress in polymer optical fiber light sources and fiber Bragg gratings," *IEEE J. Sel. Top. Quantum Electron.*, vol. 23, no. 2, 2017.

22. X. Hu, D. Kinet, K. Chah, C.-F. J. Pun, H.-Y. Tam, and C. Caucheteur, "Bragg grating inscription in PMMA optical fibers using 400-nm femtosecond pulses," *Opt. Lett.*, vol. 42, no. 14, pp. 2794–2797, 2017.

23. X. M. Tao, J. M. Yu, and H.-Y. Tam, "Photosensitive polymer optical fibres and gratings," *Trans. Inst. Meas. Control*, vol. 29, no. 3–4, pp. 255–270, 2007.

24. N. Tanio and M. Irie, "Refractive index of organic photochromic dye-amorphous polymer composites," *Jpn. J. Appl. Phys.*, vol. 33, no. 7A, pp. 3942–3946, 1994.

25. K. Kuriki, T. Kobayashi, N. Imai, T. Tamura, Y. Koike, and Y. Okamoto, "Organic dye-doped polymer optical fiber lasers," *Polym. Adv. Technol.*, vol. 11, no. 8–12, pp. 612–616, 2000.

26. M. Large, S. Ponrathnam, A. Argyros, N. Pujari, and F. Cox, "Solution doping of microstructured polymer optical fibres," *Opt. Express*, vol. 12, no. 9, pp. 1966–1971, 2004.

27. J. Arrue, F. Jiménez, I. Ayesta, M. A. Illarramendi, and J. Zubia, "Polymer-optical-fiber lasers and amplifiers doped with organic dyes," *Polymers*, vol. 3, no. 4, pp. 1162–1180, 2011.

28. P. Stajanca, I. Topolniak, S. Pötschke, and K. Krebber, "Solution-mediated cladding doping of commercial polymer optical fibers," *Opt. Fiber Technol.*, vol. 41, pp. 227–234, 2018.

29. M. C. J. Large, L. Poladian, G. W. Barton, and M. A. van Eijkelenborg, *Microstructured Polymer Optical Fibres*. Boston, MA: Springer, 2008.

30. Mitsubishi Rayon, "General Properties of AcrypetTM." 2015.

31. G. Khanarian and H. Celanese, "Optical properties of cyclic olefin copolymers," *Opt. Eng.*, vol. 40, no. 6, pp. 1024–1029, 2001.

32. K. Minami, *Handbook of Plastic Optics*, 2nd ed. Weinheim: Wiley-VCH Verlag GmbH & Co. KGaA, 2010.

33. Goodfellow Corporation, "Standard Price List for all Polymers." 2016.

34. Teijin Limited, "Panlite® AD-5503- Polycarbonate." 2016.

35. AmericasStyrenics LLC, "STYRON 666D General Purpose Polystyrene Resin." 2008.

36. Topas Advanced Polymers GmbH, "Cycloolefin Copolymer (COC)." 2013.

37. Zeon Chemicals, "ZEONEX® Cyclo Olefin Polymer (COP)." 2016.

38. Asahi Glass Co. Ltd., "Amorphous Fluoropolymer (CYTOP)." Tokyo, 2009.

39. A. Lacraz, M. Polis, A. Theodosiou, C. Koutsides, and K. Kalli, "Bragg grating inscription in CYTOP polymer optical fibre using a femtosecond laser," in *Micro-Structured and Specialty Optical Fibres IV*, K. Kalli, J. Kanka and A. Mendez, Eds. Bellingham: SPIE, 2015, vol. 9507, p. 95070K.

40. Y. Luo, B. Yan, Q. Zhang, G.-D. Peng, J. Wen, and J. Zhang, "Fabrication of polymer optical fibre (POF) gratings," *Sensors*, vol. 17, no. 511, 2017.

41. Y. Shao, R. Cao, Y.-K. Huang, P. N. Ji, and S. Zhang, "112-Gb/s transmission over 100 m of graded-index plastic optical fiber for optical data center applications," in *Optical Fiber Communication Conference*, 2012, p. OW3J.5.

42. C. Markos, W. Yuan, K. Vlachos, G. E. Town, and O. Bang, "Label-free biosensing with high sensitivity in dual-core microstructured polymer optical fibers," *Opt. Express*, vol. 19, no. 8, pp. 7790–7798, 2011.

43. K. S. C. Kuang, S. T. Quek, C. G. Koh, W. J. Cantwell, and P. J. Scully, "Plastic optical fibre sensors for structural health monitoring: A review of recent progress," *J. Sensors*, vol. 2009, 2009.

44. L. Bilro, J. G. Oliveira, J. L. Pinto, and R. N. Nogueira, "A reliable low-cost wireless and wearable gait monitoring system based on a plastic optical fibre sensor," *Meas. Sci. Technol.*, vol. 22, no. 4, p. 045801, 2011.

45. J. Zubia and J., Arrue, "Plastic optical fibers: An introduction to their technological processes and applications," *Opt. Fiber Technol.*, vol. 7, pp. 101–140, 2001.

46. O. Ziemann, J. Krauser, P. E. Zamzow, and W. Daum, *POF Handbook*. Berlin and Heidelberg: Springer, 2008.

47. W. Groh, "Overtone absorption in macromolecules for polymer optical fibers," *Macromol. Chem. Phys.*, vol. 189, no. 12, pp. 2861–2874, 1988.

48. C. Zhang, W. Zhang, D. J. Webb, and G.-D. Peng, "Optical fibre temperature and humidity sensor," *Electron. Lett.*, vol. 46, no. 9, pp. 643–644, 2010.

49. W. Zhang, D. J. Webb, and G.-D. Peng, "Polymer optical fiber Bragg grating acting as an intrinsic biochemical concentration sensor," *Opt. Lett.*, vol. 37, no. 8, pp. 1370–1372, 2012.

50. T. Kaino, M. Fujiki, and S. Nara, "Low-loss polystyrene core-optical fibers," *J. Appl. Phys.*, vol. 52, no. 12, pp. 7061–7063, 1981.

51. Y. Koike and K. Koike, "Progress in low-loss and high-bandwidth plastic optical fibers," *J. Polym. Sci. Part B Polym. Phys.*, vol. 49, no. 1, pp. 2–17, 2011.

52. C. Emslie, "Polymer optical fibres," *J. Mater. Sci.*, vol. 23, no. 7, pp. 2281–2293, 1988.
53. J. Spigulis, "Side-emitting fibers brighten our world," *Opt. Photonics News*, vol. 16, no. 10, pp. 34–39, 2005.
54. K. Obuchi, M. Komatsu, and K. Minami, "High performance optical materials cyclo olefin polymer ZEONEX," in *Optical Manufacturing and Testing VII*, J. H. Burge, O.W. Faehnle and R. Williamson, Eds. Bellingham: SPIE, 2007, vol. 6671, p. 66711I.
55. J. Cui, J. X. Yang, Y. G. Li, and Y. S. Li, "Synthesis of high performance cyclic olefin polymers (COPs) with ester group via ring-opening metathesis polymerization," *Polymers*, vol. 7, no. 8, pp. 1389–1409, 2015.
56. G. Emiliyanov et al., "Localized biosensing with Topas microstructured polymer optical fiber," *Opt. Lett.*, vol. 32, no. 5, pp. 460–462, 2007.
57. K. Nielsen, H. K. Rasmussen, A. J. Adam, P. C. Planken, O. Bang, and P. U. Jepsen, "Bendable, low-loss Topas fibers for the terahertz frequency range," *Opt. Express*, vol. 17, no. 10, pp. 8592–8601, 2009.
58. C. Markos, A. Stefani, K. Nielsen, H. K. Rasmussen, W. Yuan, and O. Bang, "High-Tg TOPAS microstructured polymer optical fiber for fiber Bragg grating strain sensing at 110 degrees," *Opt. Express*, vol. 21, no. 4, pp. 4758–4765, 2013.
59. W. Yuan et al., "Humidity insensitive TOPAS polymer fiber Bragg grating sensor," *Opt. Express*, vol. 19, no. 20, pp. 19731–19739, 2011.
60. J. Anthony, R. Leonhardt, A. Argyros, and M. C. J. Large, "Characterization of a microstructured Zeonex terahertz fiber," *J. Opt. Soc. Am. B*, vol. 28, no. 5, pp. 1013–1018, 2011.
61. G. Woyessa et al., "Humidity insensitive step-index polymer optical fibre Bragg grating sensors," in *24th International Conference on Optical Fiber Sensors*, H. J. Kalinowski, J. L. Fabris and W. J. Bock, Eds. Bellingham: SPIE, 2015, vol. 9634, p. 96342L.
62. S. G. Leon-Saval, R. Lwin, and A. Argyros, "Multicore composite single-mode polymer fiber," *Opt. Express*, vol. 20, no. 1, pp. 141–148, 2012.
63. K. Yamamoto and G. Ogawa, "Structure determination of the amorphous perfluorinated homopolymer: Poly [perfluoro(4-vinyloxyl-1-butene)]," *J. Fluor. Chem.*, vol. 126, no. 9–10, pp. 1403–1408, 2005.
64. Y. Koike and T. Ishigure, "High-bandwidth plastic optical fiber for fiber to the display," *J. Light. Technol.*, vol. 24, no. 12, pp. 4541–4553, 2006.
65. N. Tanio and Y. Koike, "What is the most transparent polymer?" *Polym. J.*, vol. 32, no. 1, pp. 43–50, 2000.
66. N. Sultanova, S. Kasarova, and I. Nikolov, "Dispersion properties of optical polymers," *Acta Phys. Pol. A.*, vol. 116, no. 4, pp. 585–587, 2009.
67. J. B. Jensen, P. E. Hoiby, and L. H. Pedersen, "Selective detection of antibodies in microstructured polymer optical fibers," *Opt. Express*, vol. 13, no. 15, pp. 5883–5889, 2005.
68. F. Berghmans et al., "Poly(D,L-lactic acid) (PDLLA) biodegradable and biocompatible polymer optical fiber," *J. Light. Technol.*, vol. 37, no. 9, pp. 1916–1923, 2019.
69. G. Rajan, K. Bhowmik, J. Xi, and G. D. Peng, "Etched polymer fibre bragg gratings and their biomedical sensing applications," *Sensors*, vol. 17, no. 10, p. 2336, 2017.
70. I.-L. Bundalo, R. Lwin, S. Leon-Saval, and A. Argyros, "All-plastic fiber-based pressure sensor," *Appl. Opt.*, vol. 55, no. 4, pp. 811–816, 2016.
71. C. Broadway et al., "Fabry-Perot micro-structured polymer optical fibre sensors for opto-acoustic endoscopy," in *Biophotonics South America*, C. Kurachi, K. Svanberg, B. J. Tromberg and V. S. Bagnato, Eds. Bellingham: SPIE, 2015, vol. 9531, p. 953116.
72. C. Broadway et al., "Microstructured polymer optical fibre sensors for opto-acoustic endoscopy," in *Micro-Structured and Specialty Optical Fibres IV*, K. Kalli and A. Mendez, Eds. Bellingham: SPIE, 2016, vol. 9886, p. 98860S.

73. D. Gallego and H. Lamela, "Microstructured polymer optical fiber sensors for optoacoustic endoscopy," in *Photons Plus Ultrasound: Imaging and Sensing*, A. A. Oraevsky and L. V. Wang, Eds. Bellingham: SPIE, 2017, vol. 10064, p. 1006412.

74. K. Krebber, P. Lenke, S. Liehr, J. Witt, and M. Schukar, "Smart technical textiles with integrated POF sensors," in *Smart Sensor Phenomena, Technology, Networks, and Systems*, W. Ecke, K. J. Peters, N. G. Meyendorf, Eds. Bellingham: SPIE, 2008, vol. 6933, p. 69330V.

75. B. C. Yao et al., "Graphene-based D-shaped polymer FBG for highly sensitive erythrocyte detection," *IEEE Photonics Technol. Lett.*, vol. 27, no. 22, pp. 2399–2402, 2015.

76. D. Vilarinho et al., "POFBG-embedded cork insole for plantar pressure monitoring," *Sensors*, vol. 17, no. 12, p. 2924, 2017.

77. A. Leal-Junior et al., "Fiber Bragg gratings in CYTOP fibers embedded in a 3D-printed flexible support for assessment of human–robot interaction forces," *Materials*, vol. 11, no. 11, p. 2305, 2018.

78. T. Kaino, "Preparation of plastic optical fibers for near-IR region transmission," *J. Polym. Sci. Part A Polym. Chem.*, vol. 25, no. 1, pp. 37–46, 1987.

79. F. Urbach, "The long-wavelength edge of photographic sensitivity and of the electronic absorption of solids," *Phys. Rev.*, vol. 92, no. 5, p. 1324, 1953.

80. T. Kaino, "Absorption losses of low loss plastic optical fibers," *Jpn. J. Appl. Phys.*, vol. 24, no. 12, pp. 1661–1665, 1985.

81. A. Gupta, R. Liang, F. D. Tsay, and J. Moacanin, "Characterization of a dissociative excited state in the solid state: Photochemistry of poly(methyl methacrylate). Photochemical processes in polymeric systems," *Macromolecules*, vol. 13, no. 6, pp. 1696–1700, 1980.

82. A. Stefani, C. Markos, and O. Bang, "Narrow bandwidth 850-nm fiber Bragg gratings in few-mode polymer optical fibers," *IEEE Photonics Technol. Lett.*, vol. 23, no. 10, pp. 660–662, 2011.

83. I. P. Johnson et al., "Polymer PCF Bragg grating sensors based on poly(methyl methacrylate) and TOPAS cyclic olefin copolymer," in *Optical Sensors 2011 and Photonic Crystal Fibers V*, 2011, vol. 8073, p. 80732V.

84. D. J. Webb, "Polymer fiber Bragg grating sensors and their applications," in *Optical Fiber Sensors: Advanced Techniques and Applications*, G. Rajan, Ed. Boca Raton, FL: CRC Press, 2015, pp. 257–276.

85. A. Argyros, "Microstructured polymer optical fibers," *J. Light. Technol.*, vol. 27, no. 11, pp. 1571–1579, 2009.

86. M. Naritomi, "'CYTOP®' amorphous fluoropolymers for low loss POF," in *POF Asia-Pacific Forum*, 1996, pp. 23–27.

87. T. Yamashita and K. Kamada, "Intrinsic transmission loss of polycarbonate core optical fiber," *Jpn. J. Appl. Phys.*, vol. 32, no. 6A, pp. 2681–2686, 1993.

88. "Paradigm the Polymer Division of Incom, Inc," 2016. [Online]. Available: www.paradigmoptics.com/. [Accessed: June 6, 2018].

89. C. Lethien, C. Loyez, J. P. Vilcot, N. Rolland, and P. A. Rolland, "Exploit the bandwidth capacities of the perfluorinated graded index polymer optical fiber for multiservices distribution," *Polymers*, vol. 3, no. 4, pp. 1006–1028, 2011.

90. G. Woyessa, A. Fasano, C. Markos, A. Stefani, H. K. Rasmussen, and O. Bang, "Zeonex microstructured polymer optical fiber: fabrication friendly fibers for high temperature and humidity insensitive Bragg grating sensing," *Opt. Mater. Express*, vol. 7, no. 1, pp. 286–295, 2017.

91. M. Rosenberger, G.-L. Roth, B. Adelmann, B. Schmauss, and R. Hellmann, "Temperature referenced planar bragg grating strain sensor in fs-Laser cut COC specimen," *IEEE Photonics Technol. Lett.*, vol. 29, no. 11, pp. 885–888, 2017.

92. R. Nogueira, R. Oliveira, L. Bilro, and J. Heidarialamdarloo, "New advances in polymer fiber Bragg gratings," *Opt. Laser Technol.*, vol. 78, pp. 104–109, 2016.

93. X. Cheng, J. Bonefacino, B. O. Juan, and H. Y. Tam, "All-polymer fiber-optic pH sensor," *Opt. Express*, vol. 26, no. 11, pp. 14610–14616, 2018.

94. G. Rajan, B. Liu, Y. Luo, E. Ambikairajah, and G.-D. Peng, "High sensitivity force and pressure measurements using etched singlemode polymer fiber Bragg gratings," *IEEE Sens. J.*, vol. 13, no. 5, pp. 1794–1800, 2013.

95. B. Yan et al., "Simultaneous vector bend and temperature sensing based on a polymer and silica optical fibre grating pair," *Sensors*, vol. 18, no. 10, p. 3507, 2018.

96. R. N. Nogueira, L. Bilro, C. A. F. Marques, R. F. Oliveira, and J. Heidarialamdarloo, "Bragg gratings in plastic optical fibre for communications and sensing applications," in *22nd International Conference on Plastical Optical Fibers, POF 2013*, 2013, pp. 25–30.

97. P. Antunes, H. Lima, J. Monteiro, and P. S. Andre, "Elastic constant measurement for standard and photosensitive single mode optical fibres," *Microw. Opt. Technol. Lett.*, vol. 50, no. 9, pp. 2467–2469, 2008.

98. D. Webb et al., "Grating and interferometric devices in POF," in *14th International Conference on Polymer Optical Fibers*, 2005, pp. 325–328.

99. C. Jiang, M. G. Kuzyk, J.-L. Ding, W. E. Johns, and D. J. Welker, "Fabrication and mechanical behavior of dye-doped polymer optical fiber," *J. Appl. Phys.*, vol. 92, no. 1, pp. 4–12, 2002.

100. W. Yuan et al., "Improved thermal and strain performance of annealed polymer optical fiber Bragg gratings," *Opt. Commun.*, vol. 284, no. 1, pp. 176–182, 2011.

101. A. Abang and D. J. Webb, "Effects of annealing, pre-tension and mounting on the hysteresis of polymer strain sensors," *Meas. Sci. Technol.*, vol. 25, no. 1, p. 015102, 2014.

102. A. Pospori et al., "Annealing effects on strain and stress sensitivity of polymer optical fibre based sensors," in *Micro-Structured and Specialty Optical Fibres IV*, K. Kalli and A. Mendez, Eds. Bellingham: SPIE, 2016, vol. 9886, p. 98860V.

103. R. Oliveira, T. H. R. Marques, L. Bilro, R. Nogueira, and C. M. B. Cordeiro, "Multiparameter POF sensing based on multimode interference and fiber Bragg grating," *J. Light. Technol.*, vol. 35, no. 1, pp. 3–9, 2017.

104. C. C. Ye et al., "Applications of polymer optical fibre grating sensors to condition monitoring of textiles," *J. Phys. Conf. Ser.*, vol. 178, no. 1, p. 012020, 2009.

105. H. Guerrero, G. V. Guinea, and J. Zoido, "Mechanical properties of polycarbonate optical fibers," *Fiber Integr. Opt.*, vol. 17, no. 3, pp. 231–242, 1998.

106. S. Liehr, P. Lenke, and M. Wendt, "Polymer optical fiber sensors for distributed strain measurement and application in structural health monitoring," *IEEE Sens. J.*, vol. 9, no. 11, pp. 1330–1338, 2009.

107. S. Atakaramians et al., "Cleaving of extremely porous polymer fibers," *IEEE Photonics J.*, vol. 1, no. 6, pp. 286–292, 2009.

108. J. Canning, E. Buckley, N. Groothoff, B. Luther-Davies, and J. Zagari, "UV laser cleaving of air–polymer structured fibre," *Opt. Commun.*, vol. 202, no. 1–3, pp. 139–143, 2002.

109. O. Abdi, K. C. Wong, T. Hassan, K. J. Peters, and M. J. Kowalsky, "Cleaving of solid single mode polymer optical fiber for strain sensor applications," *Opt. Commun.*, vol. 282, no. 5, pp. 856–861, 2009.

110. S. H. Law et al., "Cleaving of microstructured polymer optical fibres," *Opt. Commun.*, vol. 258, no. 2, pp. 193–202, 2006.

111. A. Stefani, K. Nielsen, H. K. Rasmussen, and O. Bang, "Cleaving of TOPAS and PMMA microstructured polymer optical fibers: Core-shift and statistical quality optimization," *Opt. Commun.*, vol. 285, no. 7, pp. 1825–1833, 2012.

112. S. H. Law, M. A. van Eijkelenborg, G. W. Barton, C. Yan, R. Lwin, and J. Gan, "Cleaved end-face quality of microstructured polymer optical fibres," *Opt. Commun.*, vol. 265, no. 2, pp. 513–520, 2006.
113. A. Abang and D. J. Webb, "Demountable connection for polymer optical fiber grating sensors," *Opt. Eng.*, vol. 51, no. 8, p. 080503, 2012.
114. A. Abang, D. Saez-Rodriguez, K. Nielsen, O. Bang, and D. J. Webb, "Connectorisation of fibre Bragg grating sensors recorded in microstructured polymer optical fibre," in *Fifth European Workshop on Optical Fibre Sensors (EWOFS'2013)*, L. R. Jaroszewicz, Ed. Bellingham: SPIE, 2013, vol. 8794, p. 87943Q.
115. R. Oliveira, L. Bilro, and R. Nogueira, "Smooth end face termination of microstructured, graded-index, and step-index polymer optical fibers," *Appl. Opt.*, vol. 54, no. 18, pp. 5629–5633, 2015.
116. J. Witt, M. Breithaupt, J. Erdmann, and K. Krebber, "Humidity sensing based on microstructured POF long period gratings," in *20th International Conference on Plastic Optical Fibers*, 2011, pp. 409–414.
117. M. V. P. Ghirghi, V. Minkovich, and A. G. Villegas, "Polymer optical fiber termination with use of liquid nitrogen," *IEEE Photonics Technol. Lett.*, vol. 26, no. 5, pp. 516–519, 2014.
118. R. Lwin and A. Argyros, "Connecting microstructured polymer optical fibres to the world," in *18th International Conference on Plastic Optical Fibers*, 2009, vol. 18, p. Poster 7.
119. D. Sáez-Rodriguez, K. Nielsen, O. Bang, and D. Webb, "Simple room temperature method for polymer optical fibre cleaving," *J. Light. Technol.*, vol. 33, no. 23, pp. 4712–4716, 2015.
120. D. Saez-Rodriguez, R. Min, B. Ortega, K. Nielsen, and D. J. Webb, "Passive and portable polymer optical fiber cleaver," *IEEE Photonics Technol. Lett.*, vol. 28, no. 24, pp. 2834–2837, 2016.
121. D. Sáez-Rodríguez, K. Nielsen, O. Bang, and D. J. Webb, "Time-dependent variation of fibre Bragg grating reflectivity in PMMA based polymer optical fibres," *Opt. Lett.*, vol. 40, no. 7, pp. 1476–1479, 2015.
122. D. Merchant, P. Scully, and N. Schmitt, "Chemical tapering of polymer optical fibre," *Sensors Actuators A Phys.*, vol. 76, no. 1–3, pp. 365–371, 1999.
123. K. Bhowmik et al., "Experimental study and analysis of hydrostatic pressure sensitivity of polymer fibre Bragg gratings," *J. Light. Technol.*, vol. 33, no. 12, pp. 2456–2462, 2015.
124. R. Inglev, J. Janting, K. Nielsen, G. Woyessa, and O. Bang, "The application of hansen solubility parameters for local etching of TOPAS polymer optical fibers," in *26th International Conference on Optical Fiber Sensors*, L. Thévenaz et al., Eds. Washington, DC: OSA, 2018, p. TuE66.
125. R. Gravina, G. Testa, and R. Bernini, "Perfluorinated plastic optical fiber tapers for evanescent wave sensing," *Sensors*, vol. 9, no. 12, pp. 10423–10433, 2009.
126. D. Sáez-Rodríguez, K. Nielsen, H. K. Rasmussen, O. Bang, and D. J. Webb, "Highly photosensitive polymethyl methacrylate microstructured polymer optical fiber with doped core," *Opt. Lett.*, vol. 38, no. 19, pp. 3769–3772, 2013.
127. L. Pereira et al., "Polymer optical fiber Bragg grating inscription with a single Nd:YAG laser pulse," *Opt. Express*, vol. 26, no. 14, pp. 18096–18104, 2018.
128. M. C. Zanon, V. N. H. Silva, A. P. L. Barbero, and R. M. Ribeiro, "Practical splicing of poly-methyl-methacrylate plastic optical fibers," *Appl. Opt.*, vol. 57, no. 4, pp. 812–816, 2018.
129. A. Argyros, M. A. van Eijkelenborg, S. D. Jackson, and R. P. Mildren, "Microstructured polymer fiber laser," *Opt. Lett.*, vol. 29, no. 16, pp. 1882–1884, 2004.

130. M. K. Szczurowski, T. Martynkien, G. Statkiewicz-Barabach, W. Urbanczyk, and D. J. Webb, "Measurements of polarimetric sensitivity to hydrostatic pressure, strain and temperature in birefringent dual-core microstructured polymer fiber," *Opt. Express*, vol. 18, no. 12, pp. 12076–12087, 2010.

131. M. Ferreira, A. Gomes, D. Kowal, G. Statkiewicz-Barabach, P. Mergo, and O. Frazão, "The fiber connection method using a tapered silica fiber tip for microstructured polymer optical fibers," *Fibers*, vol. 6, no. 1, p. 4, 2018.

132. X. Hu, C.-F. J. Pun, H.-Y. Tam, P. Mégret, and C. Caucheteur, "Highly reflective Bragg gratings in slightly etched step-index polymer optical fiber," *Opt. Express*, vol. 22, no. 15, pp. 18807–18817, 2014.

133. I. P. Johnson, D. J. Webb, K. Kalli, M. C. Large, and A. Argyros, "Multiplexed FBG sensor recorded in multimode microstructured polymer optical fibre," in *Photonic Crystal Fibres IV*, K. Kalli and W. Urbanczyk, Eds. Bellingham: SPIE, 2010, vol. 7714, p. 77140D.

134. D. Marcuse, "Loss analysis of single-mode fiber splices," *Bell Syst. Tech. J.*, vol. 56, no. 5, pp. 703–718, 1977.

135. H. Kogelnik, "Coupling and conversion coefficients for optical modes in quasi-optics," in *Microwave Research Institute Symposia Series*. New York: Polytechnic Press, 1964, vol. 14, pp. 333–347.

136. M. D. Nielsen, N. A. Mortensen, J. R. Folkenberg, and A. Bjarklev, "Mode-field radius of photonic crystal fibers expressed by the V parameter," *Opt. Lett.*, vol. 28, no. 23, pp. 2309–2311, 2003.

137. N. A. Mortensen, J. R. Folkenberg, M. D. Nielsen, and K. P. Hansen, "Modal cutoff and the V parameter in photonic crystal fibers," *Opt. Lett.*, vol. 28, no. 20, pp. 1879–1881, 2003.

138. S. Nemoto and T. Makimoto, "Analysis of splice loss in single-mode fibres using a Gaussian field approximation," *Opt. Quantum Electron.*, vol. 11, no. 5, pp. 447–457, 1979.

139. M. S. Chychłowski, S. Ertman, and T. R. Woliński, "Chemical and photo-chemical bonding of polymer and silica fibers," *Photonics Lett. Pol.*, vol. 6, no. 3, pp. 111–113, 2014.

140. X. Hu et al., "Polarization effects in polymer FBGs: Study and use for transverse force sensing," *Opt. Express*, vol. 23, no. 4, pp. 4581–4590, 2015.

141. D. Saez-Rodriguez, J. L. Cruz, I. Johnson, D. J. Webb, M. C. J. Large, and A. Argyros, "Water diffusion into UV inscripted long period grating in microstructured polymer fiber," *IEEE Sens. J.*, vol. 10, no. 7, pp. 1169–1173, 2010.

142. I. Johnson, *Grating Devices in Polymer Optical Fibre*. UK: Aston University, 2011.

143. C. Zhang, *Fibre Bragg Gratings in Polymer Optical Fibre for Applications in Sensing rating Devices in Polymer Optical Fibre*. UK: Aston University, 2011.

144. Z. F. Zhang, C. Zhang, X. M. Tao, G. F. Wang, and G. D. Peng, "Inscription of polymer optical fiber Bragg grating at 962 nm and its potential in strain sensing," *IEEE Photonics Technol. Lett.*, vol. 22, no. 21, pp. 1562–1564, 2010.

145. X. Chen, W. Zhang, C. Liu, Y. Hong, and D. J. Webb, "Enhancing the humidity response time of polymer optical fiber Bragg grating by using laser micromachining," *Opt. Express*, vol. 23, no. 20, pp. 25942–25949, 2015.

146. A. Stefani, S. Andresen, W. Yuan, N. Herholdt-Rasmussen, and O. Bang, "High sensitivity polymer optical fiber-bragg-grating-based accelerometer," *IEEE Photonics Technol. Lett.*, vol. 24, no. 9, pp. 763–765, 2012.

147. X. Hu, P. Mégret, and C. Caucheteur, "Surface plasmon excitation at near-infrared wavelengths in polymer optical fibers," *Opt. Lett.*, vol. 40, no. 17, pp. 3998–4001, 2015.

148. R. Oliveira, T. H. R. Marques, and C. M. B. Cordeiro, "Strain sensitivity enhancement of a sensing head based on ZEONEX polymer FBG in series with silica fiber," *J. Light. Technol.*, vol. 36, no. 22, pp. 5106–5112, 2018.

149. D. Sáez-Rodríguez, J. L. C. Munoz, I. Johnson, D. J. Webb, M. C. J. Large, and A. Argyros, "Long period fibre gratings photoinscribed in a microstructured polymer optical fibre by UV radiation," in *Photonic Crystal Fibers III*, K. Kalli, Ed. Bellingham: SPIE, 2009, vol. 7357, p. 73570L.

150. B. Van Hoe et al., "Optical fiber sensors embedded in flexible polymer foils," in *Optical Sensing and Detection*, F. Berghmans, A. G. Mignani and C. A. van Hoof, Eds. Bellingham: SPIE, 2010, vol. 7726, p. 772603.

151. Norland, "Norland UV Curing Optical Adhesives; Technical Data," 2016.

152. Ellsworth Adhesives, "The Adhesive Sourcebook," 2015.

153. R. Oliveira, L. Bilro, and R. Nogueira, "Strain sensitivity control of an in-series silica and polymer FBG," *Sensors*, vol. 18, no. 6, p. 1884, Jun. 2018.

154. R. Oliveira, L. Bilro, and R. Nogueira, "Strain and temperature detection through PFBG and resin based FP cavities," in *26th International Conference on Optical Fiber Sensors*, L. Thévenaz et al., Eds. Washington, DC: OSA, 2018, p. WF78.

155. R. Oliveira, L. Bilro, and R. Nogueira, "Simultaneous measurement of temperature and humidity using PFBG and Fabry-Perot cavity," in *26th International Conference on Plastic Optical Fibres*, 2017, p. Paper 83.

156. R. Oliveira, L. Bilro, T. H. R. Marques, C. M. B. Cordeiro, and R. Nogueira, "Simultaneous detection of humidity and temperature through an adhesive based Fabry – Pérot cavity combined with polymer fiber Bragg grating," *Opt. Lasers Eng.*, vol. 114, pp. 37–43, 2019.

2 Fiber Bragg Grating Fabrication

2.1 INTRODUCTION

Fiber Bragg grating (FBG) is a device based on a periodic modulation of the refractive index along the length of an optical fiber and is produced through the exposure of the fiber core to an intense optical interference pattern. This periodic structure acts as a selective mirror for the wavelength that satisfies the Bragg condition expressed as:

$$\lambda_{\text{Bragg}} = 2n_{\text{eff}}\Lambda \tag{2.1}$$

where Λ is the grating period and n_{eff} is the effective refractive index of the guided mode. The first inscription of a permanent grating in an optical fiber was demonstrated by Hill et al. in 1978 [1], by launching an intense Argon-ion laser into an optical fiber doped with germanium, forming a standing-wave pattern that resulted in a periodic refractive index modification of the core optical fiber. Since then, the interest in these grating structures has increased and different inscription techniques have been developed [2]. Nowadays, FBGs are commercially available and have been applied extensively in different systems and sub-systems targeting applications in communications and sensors.

In 1999, the group of Peng et al. motivated by the high elongation capabilities offered by polymers, reported the inscription of polymer optical fiber Bragg gratings (POFBGs) in step-index (SI) fibers, either multimode (MM) [3] or single-mode (SM) [4], at the infrared region, using a 325 nm UV radiation. In 2005, four years after the production of the first microstructured polymer optical fibers (mPOF) [5], Dobb et al. reported the inscription of Bragg gratings at the infrared region in mPOFs, either SM or MM [6], using the same wavelength radiation. POFBGs have been essentially written in polymethylmethacrylate (PMMA) optical fibers due to the wide availability of this polymer material. Nevertheless, POFBGs in other polymers such as TOPAS® [7], polycarbonate (PC) [8] polystyrene (PS) [9], ZEONEX® [10,11] and also cyclic transparent optical polymer (CYTOPR®) [12,13], have also been reported. Despite the high attenuation for the majority of POFs at the infrared region, most of the POFBGs have been described at this spectral window. The reason is mainly due to the availability of sources and detectors for telecom applications, developed for the widely used silica optical fiber, which has low loss at this spectral region. Nevertheless, the inscription of POFBGs has also been demonstrated in different wavelength regions, including the low loss region of polymers, located at the visible region [14,15]. The writing methodology has fundamentally been done through the phase mask technique [6,15,16], due to the easy implementation of the recording setup. On the other hand, other inscription methods, such as the combination of the phase mask and ring interferometry [4,17,18], and the point-by-point (PbP) and line-by-line (LbL) through a femtosecond (fs) laser [13,19], have also been used.

This chapter will focus mainly on the production of fiber Bragg gratings in POFs made of different materials and structures. Different laser sources have been reported along the years for the recording of grating structures in polymer materials. A detailed description of those works is not the scope of this chapter, instead, it will focus on the lasers having the most efficient inscription wavelength, namely the 325 nm continuous-wave (CW) Helium-Cadmium (HeCd) laser and the 248 nm Krypton-Fluoride (KrF) pulsed laser. The different breakthroughs achieved along the years, namely by the use of different laser sources for the reduction of the inscription time, the use of doped fibers and the influence of the microstructured fiber orientation on the irradiation process are key aspects for this technology and therefore, they will be under analysis in this chapter. Nevertheless, the inscription of POFBGs at the low loss region of polymers brought the possibility to extend the technology a step forward and thus, the methodologies used so far will be reviewed. Finally, the recent achievements regarding the Bragg grating inscription through fs lasers will be described.

2.2 PHOTOSENSITIVITY IN POLYMER MATERIALS

Photosensitivity is the term used in the context of writing waveguides or Bragg gratings in optical materials and it describes the induced change in the material optical properties upon irradiation. The description of photosensitivity in PMMA, the most known polymer, can be traced back to the early 1970s, well before the discovery of photosensitivity in silica fibers. These works were done at the Bell Laboratories using a 325 nm radiation, being this the most preferred UV source during these days. Tomlinson et al. [20] were pioneered in one of those works, where thin PMMA slabs were exposed to a 325 nm HeCd laser, showing a significant increase in refractive index (3×10^{-3}). The capability to produce holographic gratings by a two-beam interference process was also demonstrated, achieving efficiencies as large as 70% with resolutions of ~5,000 lines/mm. In this work, they assumed that the increasing refractive index was due to the crosslinking through the oxidation products of the monomer. Consequently, the polymer chains became closer, leading to an increase in density and hence, an increase of the refractive index. However, in 1974, studies developed by Bowden et al. [21] revealed no evidence for this description. They concluded that the mechanism involved in the refractive index increase in the irradiated region was due to the photoinduced polymerization of the residual monomers that did not react in the polymerization process (1%–2%), resulting in an increase of molecular density. The density change is the result of the bonding between the monomer molecules, leading to shorter intermolecular distances, which promote densification and hence, refractive index increase. By judging the previous conclusions, one may think of using polymer materials with a high rate of unreacted monomers, in order to increase the refractive index change upon irradiation. However, such excess monomers will tend to polymerize over time, leading to the formation of gases and inherently bubbles in the polymer material, causing scattering of light and thus, raising the attenuation of the material [22].

In 1984, a work by Kopietz and co-workers [23] reported the irradiation of PMMA with a low-pressure mercury lamp and observed a decrease of the refractive index in the first stage of illumination, followed by an increase until saturation. The process was explained as an increase of the presence of new monomers under UV radiation, followed by a second stage where a photopolymerization of those monomers prevailed until the saturation of the refractive index. However, the authors did not explain the chemical reactions involved in the process. In 1990 [24] and 1993 [25], it was demonstrated that photodegradation occurs at wavelengths below 320 nm, leading to the formation of free radicals and monomers. In such work, the authors conclude that under 320 nm wavelength radiation, the primary photochemical reaction was the direct main chain scission of the polymer (see Figure 2.1), forming radicals and monomers. The results were similar to the samples irradiated in air and vacuum.

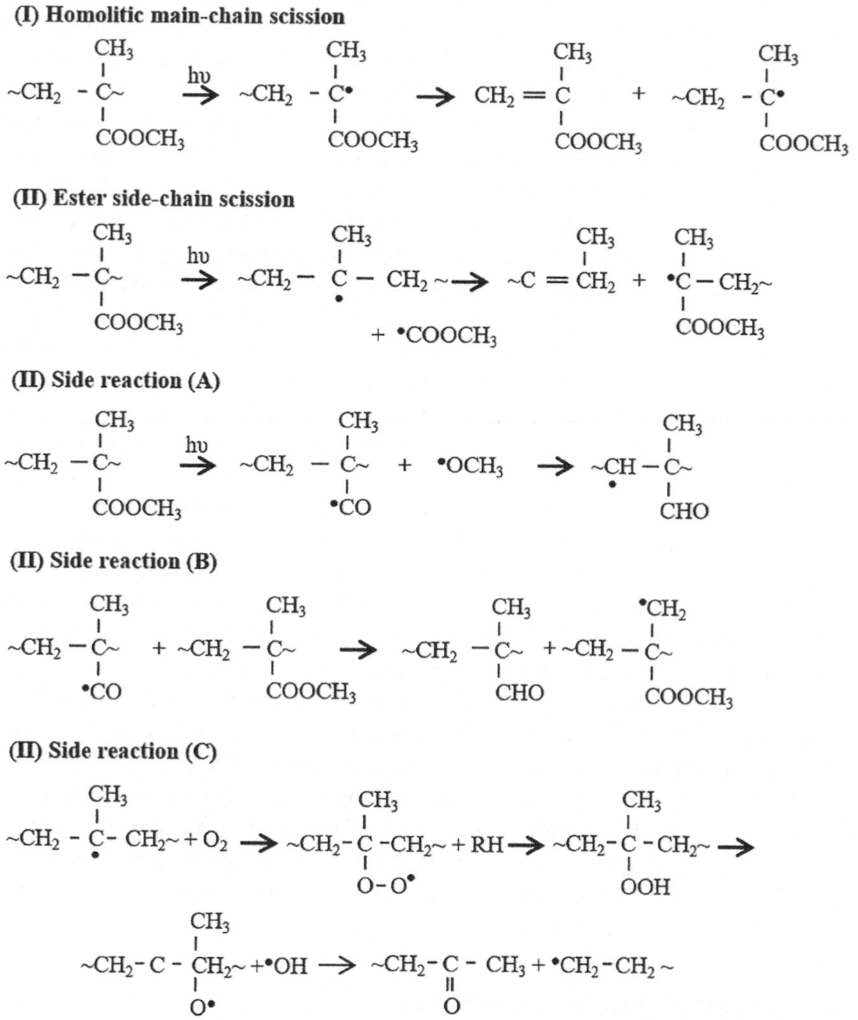

(I) Homolitic main-chain scission

$$\sim CH_2 - \underset{\underset{COOCH_3}{|}}{\overset{\overset{CH_3}{|}}{C}}\sim \;\xrightarrow{h\upsilon}\; \sim CH_2 - \underset{\underset{COOCH_3}{|}}{\overset{\overset{CH_3}{|}}{C^\bullet}} \;\rightarrow\; CH_2 = \underset{\underset{COOCH_3}{|}}{\overset{\overset{CH_3}{|}}{C}} \;+\; \sim CH_2 - \underset{\underset{COOCH_3}{|}}{\overset{\overset{CH_3}{|}}{C^\bullet}}$$

(II) Ester side-chain scission

$$\sim CH_2 - \underset{\underset{COOCH_3}{|}}{\overset{\overset{CH_3}{|}}{C}}\sim \;\xrightarrow{h\upsilon}\; \sim CH_2 - \underset{\bullet}{\overset{\overset{CH_3}{|}}{C}} - CH_2 \sim \;\rightarrow\; \sim C = CH_2 \;+\; \underset{\underset{COOCH_3}{|}}{\overset{\overset{CH_3}{|}}{{}^\bullet C}} - CH_2 \sim$$

$$+\; {}^\bullet COOCH_3$$

(II) Side reaction (A)

$$\sim CH_2 - \underset{\underset{COOCH_3}{|}}{\overset{\overset{CH_3}{|}}{C}}\sim \;\xrightarrow{h\upsilon}\; \sim CH_2 - \underset{\underset{{}^\bullet CO}{|}}{\overset{\overset{CH_3}{|}}{C}}\sim \;+\; {}^\bullet OCH_3 \;\rightarrow\; \sim \underset{\bullet}{CH} - \underset{\underset{CHO}{|}}{\overset{\overset{CH_3}{|}}{C}}\sim$$

(II) Side reaction (B)

$$\sim CH_2 - \underset{\underset{{}^\bullet CO}{|}}{\overset{\overset{CH_3}{|}}{C}}\sim \;+\; \sim CH_2 - \underset{\underset{COOCH_3}{|}}{\overset{\overset{CH_3}{|}}{C}}\sim \;\rightarrow\; \sim CH_2 - \underset{\underset{CHO}{|}}{\overset{\overset{CH_3}{|}}{C}}\sim \;+ \sim CH_2 - \underset{\underset{COOCH_3}{|}}{\overset{\overset{{}^\bullet CH_2}{|}}{C}}\sim$$

(II) Side reaction (C)

$$\sim CH_2 - \underset{\bullet}{\overset{\overset{CH_3}{|}}{C}} - CH_2 \sim + O_2 \;\rightarrow\; \sim CH_2 - \underset{\underset{O-O^\bullet}{|}}{\overset{\overset{CH_3}{|}}{C}} - CH_2 \sim + RH \;\rightarrow\; \sim CH_2 - \underset{\underset{OOH}{|}}{\overset{\overset{CH_3}{|}}{C}} - CH_2 \sim \;\rightarrow$$

$$\sim CH_2 - C - \underset{\underset{O^\bullet}{|}}{\overset{\overset{CH_3}{|}}{C}}H_2 \sim + {}^\bullet OH \;\rightarrow\; \sim CH_2 - \underset{\underset{O}{\|}}{C} - CH_3 + {}^\bullet CH_2 - CH_2 \sim$$

FIGURE 2.1 Types of photodegradation process. Reaction II has three possible side reactions, being (C), formed only in the presence of oxygen. (Adapted from T. Mitsuoka, et al. "Wavelength sensitivity of the photodegradation of poly (methyl methacrylate)," *J. Appl. Polym. Sci.*, vol. 47, pp. 1027–1032, 1993)

However, for wavelengths below 320 nm, they observed the formation of other radicals, where the ester side chain scission took place, with possible side reactions ("A") and ("B"), when the samples were irradiated in vacuum, and ("C") when the samples were irradiated in air (see Figure 2.1).

The irradiation of PMMA under other wavelengths was also subject of study by the scientific community. Küper and Stuke [26] performed infrared and UV spectroscopic measurements of PMMA films irradiated with 248 nm UV source. Results allowed the explanation of an incubation process for fluences that were well below the reported for the PMMA ablation threshold [27]. Using low repetition rates ($R < 6$ Hz) and low fluence ($I = 35$–40 mJ/cm^2), the UV and infrared spectra presented new absorptions bands (ester bands), concluding that there was formation of photoproducts in the exposed areas, including methyl formate. These results were confirmed by a subsequent study done by Srinivasan et al. [28].

Baker et al. [29], in a study based on the work done by Küper and Stuke, showed a small but significant change in the refractive index of a PMMA film using a KrF laser irradiation with low repetition rate and low fluence ($R < 5$ Hz and $I = 40$ mJ/cm^2), allowing to minimize the ablative removal of PMMA. The refractive index change (Δn_{dc}) increased with the increasing number of pulses, presenting a value of 10^{-2} after 2,000 pulses [29]. Another detailed work regarding the refractive index modification with 248 nm laser radiation at low fluence ($I = 17$ mJ/cm^2 with $R = 5$ Hz), was presented by Wochnowski et al. [30]. In this work, it was concluded that the refractive index change was due to a Norrish Type I photochemical reaction. This means that a complete separation of the side chain from the PMMA molecule occurs, promoting a volume contraction by the Van der Waals interactions and consequently increase of the refractive index change [30]. Continuous irradiation at a fluence of 30 mJ/cm^2 or higher, revealed a degradation of the free ester radical, either into methane and carbon monoxide or into carbon monoxide and methanol. The complete separation of the side chain of PMMA caused the formation of a radical electron at the α-C-atom. This caused destabilization of the polymer main chain, leading to its scission at high irradiation doses and double bonding formation. This result justified the decrease of the refractive index of the polymer sample at higher doses [30].

The use of femtosecond laser has also been explored for refractive index modification of PMMA [31,32]. When a high-intensity femtosecond laser is focused on a PMMA slab, a multiphoton absorption process occurs. The mechanisms behind the photosensitivity were reported as local melting which creates density changes in the irradiated zones. Additionally, Baum et al. [33] suggest the occurrence of photochemical modifications involving direct cleavage of the polymer backbone, promoting the formation of monomers and leading to a refractive index change, measured as 4×10^{-3} on the irradiated area.

2.3   UNIFORM BRAGG GRATINGS

In its simplest form, when an optical fiber is exposed to a periodic pattern of ultraviolet radiation, a sinusoidal refractive index variation over a defined length (L_{Bragg}), with a constant period (Λ), is formed, as illustrated in Figure 2.2. The perturbation of the effective refractive index (n_{eff}) can be described as [34]:

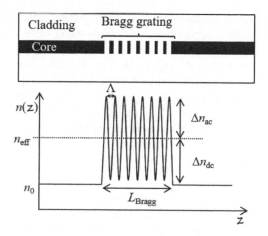

FIGURE 2.2 Refractive index modulation of a fiber Bragg grating. Δn_{dc} is the average grating index and Δn_{ac} is the grating index amplitude.

$$\Delta n_{core}(z) = n(z) - n_0 = \Delta n_{dc}(z) + \Delta n_{ac}(z)\cos\left(\frac{2\pi z}{\Lambda} + \varphi(z)\right)$$ (2.2)

where z is the position, n_0 the refractive index prior to the grating inscription, Δn_{ac} the refractive index modulation amplitude, Δn_{dc}, the average change in refractive index and $\varphi(z)$ describes the grating chirp along z, (for uniform gratings $\varphi(z) = 0$), (see Figure 2.2). The refractive index modulation amplitude remains sinusoidal until the exposed region reaches the saturation. At this point, the modulation starts to change from sinusoidal to rectangular.

Fiber Bragg gratings are simply an optical diffraction grating, and thus, its effect upon a light wave incident on the grating at an angle θ_1 can be described by the familiar grating equation:

$$n\sin(\theta_2) = n\sin(\theta_1) + m\frac{\lambda}{\Lambda}$$ (2.3)

where θ_2 is the angle of the diffracted wave and the integer m defines the diffraction order (see Figure 2.3).

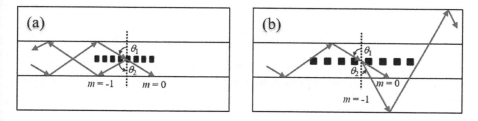

FIGURE 2.3 Ray optics illustration of the coupling of the propagating core mode to the: counter-propagating core mode (a) and co-propagating cladding mode (b).

Bragg gratings can be classified in two main groups: the ones having a short period, where the coupling occurs between modes traveling in opposite directions; and the ones where the coupling occurs between modes traveling in the same direction, also known as long-period gratings (LPGs). For the first case, illustrated in Figure 2.3a, the mode propagating with a bounce angle θ_1 is reflected into the same mode in the opposite direction with an angle $(\theta_2 = -\theta_1)$. Using the mode propagation constant defined as $\beta = n_{\text{eff}} \cdot 2\pi/\Lambda$, where the effective refractive index is defined as $n_{\text{eff}} = n_{\text{co}} \cdot \sin(\theta)$, Equation 2.3 can be rewritten as[34]:

$$\pm\beta_2 = \beta_1 + m\frac{2\pi}{\Lambda} \tag{2.4}$$

where positive values of β_2 are obtained when the coupled mode is forward, and negative values of β, when it is backward. If the first-order mode $m = -1$ is dominant, and recognizing $(\beta_2 < 0)$, there is coupling between the fundamental forward mode $n_{\text{eff},1}$ and a backward mode $n_{\text{eff},2}$ which is defined following Equation 2.4, as:

$$\lambda = \left(n_{\text{eff},1} + n_{\text{eff},2}\right)\Lambda \tag{2.5}$$

If the two modes are identical, then the familiar result for Bragg reflection is obtained as expressed in Equation 2.1.

Regarding the case shown in Figure 2.3b, the core mode propagating with bounce angle θ_1 is coupled to a co-propagating cladding mode, with a bounce angle θ_2. Since $\beta_2 > 0$, the resonance wavelength for an LPG, following Equation 2.4, is:

$$\lambda = \left(n_{\text{eff},1} - n_{\text{eff},2}\right)\Lambda \tag{2.6}$$

For co-propagating coupling at a given wavelength, evidently, a much longer grating period is required than for counter-propagating coupling.

Coupled mode theory is a good tool to obtain quantitative information about the diffraction efficiency and spectral response of fiber Bragg gratings [34]. Therefore, the reflectivity of a grating with a refractive index as expressed in Equation 2.2 and considering a uniform Δn_{ac} and Δn_{dc} can be expressed as [34]:

$$R = \tanh^2\left(\kappa L_{\text{Bragg}}\right) \tag{2.7}$$

where κ is the coupling coefficient defined as:

$$\kappa = \frac{\pi\Delta n_{\text{ac}}}{\lambda} \tag{2.8}$$

Considering the grating index changes Δn_{dc}, and the grating index amplitude Δn_{ac} shown in Figure 2.2, they can be calculated from the measured spectra using the following relations:

$$\Delta n_{\text{dc}} = \frac{n_{\text{eff}}\Delta\lambda_{\text{Bragg}}}{\lambda_{\text{Bragg}}} \tag{2.9}$$

$$\Delta n_{\text{ac}} = \frac{\lambda_{\text{Bragg}}}{\pi L_{\text{Bragg}}} \tanh^{-1}\left(\sqrt{r_{\text{max}}}\right) \tag{2.10}$$

where $n_{\text{eff}} = n_0 + \Delta n_{\text{dc}}$, and $\Delta\lambda_{\text{Bragg}}$ is the wavelength shift during the grating inscription. The theoretical grating bandwidth ($\Delta\lambda_{\text{BW}}$) can also be deduced from the coupling mode theory and it is defined as the wavelength range between the first zeros apart from the Bragg peak, and it is given by [34]:

$$\frac{\Delta\lambda_{\text{BW}}}{\lambda_{\text{Bragg}}} = \frac{\Delta n_{\text{ac}}}{n_{\text{eff}}}\sqrt{1 + \left(\frac{\lambda_{\text{Bragg}}}{\Delta n_{\text{ac}} L_{\text{Bragg}}}\right)^2} \tag{2.11}$$

Considering the "weak-grating limit" for which Δn_{ac} is very small, the above expression can be written as:

$$\frac{\Delta\lambda_{\text{BW}}}{\lambda_{\text{Bragg}}} = \frac{\lambda_{\text{Bragg}}}{n_{\text{eff}} L_{\text{Bragg}}} = \frac{2}{N}, \quad \Delta n_{\text{ac}} \ll \frac{\lambda_{\text{Bragg}}}{L_{\text{Bragg}}} \tag{2.12}$$

where N is the total number of grating periods defined as L_{Bragg}/Λ. On the other hand, for the "strong-grating limit," the expression becomes:

$$\frac{\Delta\lambda_{\text{BW}}}{\lambda_{\text{Bragg}}} = \frac{\Delta n_{\text{ac}}}{n_{\text{eff}}}, \quad \Delta n_{\text{ac}} \gg \frac{\lambda_{\text{Bragg}}}{L_{\text{Bragg}}} \tag{2.13}$$

2.4 FIBER BRAGG GRATINGS FABRICATION TECHNIQUES

Since the demonstration of the inscription of a Bragg grating in an optical fiber by Hill et al. in 1978 [1], different inscription techniques have been developed and are now widely used. These techniques are capable to offer Bragg reflection wavelengths different from the writing wavelength, as opposed to the Hill gratings. They are mostly based on the side exposure of a UV pattern to a short length of an optical fiber. In general, the side exposure techniques may be split into two general categories: the ones based on holographic methods and those that are non-interferometric, based on a simple periodic exposure of the fiber.

2.4.1 AMPLITUDE SPLITTING METHOD

The interferometric method for side inscription of FBGs was first demonstrated by Meltz et al. in 1989 [35]. In an interferometric process, a collimated laser beam is split and then recombined at a selected angle. This angle is designated to give the desired frequency of interference fringes in the target where the two beams overlap. The split of the beam is performed by a beam splitter and the resulting beams are directed to mirrors which are then reflected with the desired angle, in such a way that they interfere at a point where the sample/fiber is mounted [36]. The schematic of the technique can be seen in Figure 2.4, for the inscription of a Bragg grating in an optical fiber.

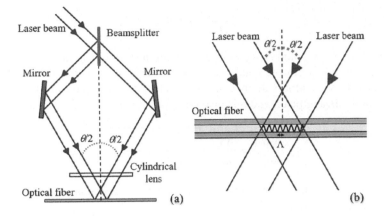

FIGURE 2.4 (a) Schematic of a two-beam interferometric arrangement used to write a uniform period grating and (b) inset of the two-beam interference on the optical fiber.

The periodicity of the interference pattern on the optical fiber leads to the formation of a Bragg reflection wavelength expressed as [37]:

$$\lambda_{Bragg} = \frac{n_{eff}\lambda_{UV}}{n_{UV}\sin\left(\dfrac{\theta}{2}\right)} \tag{2.14}$$

where n_{eff} is the effective mode index in the fiber, n_{UV} is the refractive index of the material in the ultraviolet, λ_{UV} is the wavelength of the writing radiation and θ is the mutual angle of the ultraviolet beams. From this equation, assuming that $n_{UV} \sim n_{eff}$, it is easy to understand that the Bragg wavelength can be adjusted from the one close to the inscription wavelength, until infinity when $\theta = 0$.

The construction of an interferometer as shown in Figure 2.4, requires a sturdy base and stable optical mounts. Due to the versatility of the technique, it is ideal for writing short gratings. As drawbacks, this technique is problematic for long exposures, because of the lack of mechanical stability. Consequently, it is more suited for single-pulse writing. Additionally, the path of the light beams in air should be short and the entire setup needs to be shielded from turbulence since the interference fringes formed at the fiber can drift if the paths of the two beams change during the inscription time (e.g., airflow). The production of the same interference pattern, each time the system is reconfigured is also problematic. Nevertheless, for the use of low coherence sources, the path difference between the two interfering beams must be equalized, thus, the introduction of an additional phase plate in one arm is generally required.

2.4.2 PHASE MASK METHOD

The use of a phase mask to produce a fiber Bragg grating was first demonstrated in 1993 [38]. Nowadays, it is widely used due to repeatability, simple integration, easy alignment, and the possibility of lowering the requirements of the writing source in terms of temporal and spatial coherence. A phase mask is a relief grating etched in a fused silica plate that can be fabricated using e-beam process, holographically or by femtosecond irradiation. In the phase mask inscription method, the UV beam passes through the phase mask, which diffracts the beam into several orders, $m = 0, \pm1, \pm2\ldots$. The incident and diffracted orders satisfy the general diffraction equation, with the period Λ_{pm} of the phase mask expressed as [37]:

$$\Lambda_{pm} = \frac{m\lambda_{UV}}{\sin\left(\dfrac{\theta_m}{2}\right) - \sin(\theta_i)} \tag{2.15}$$

where $\theta_m/2$ is the angle of the diffracted order, λ_{UV} the writing wavelength, and θ_i the angle of the incident UV beam. At normal incidence (i.e., $(\theta_i = 0)$, when the beam is perpendicular to the phase mask), the UV radiation is split into $m = 0$ and ±1 orders, and the interference of these two orders produces a sinusoidal intensity pattern that is used to modulate the refractive index of the fiber immediately behind the phase mask. The schematic of the phase mask inscription process can be seen in Figure 2.5.

The interference pattern of the ±1 orders brought together by parallel mirrors, as shown in Figure 2.4, has a period Λ, related to the diffraction angle $\theta_m/2$, by:

$$\Lambda = \frac{\lambda_{UV}}{2\sin\left(\dfrac{\theta_m}{2}\right)} = \frac{\Lambda_{pm}}{2} \tag{2.16}$$

where the period in the fiber core required for the Bragg wavelength defines the period of the grating etched on the phase mask, and using Equation 2.16, gives:

$$\Lambda = \frac{N\lambda_{Bragg}}{2n_{eff}} = \frac{\Lambda_{pm}}{2}, \quad N \geq 1 \tag{2.17}$$

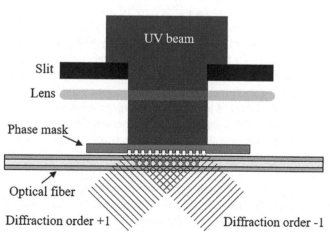

FIGURE 2.5 Schematic of the fiber Bragg grating fabrication using the phase mask technique, through the ±1 orders.

where N is the integer indicating the order of the grating. Phase masks are designed to equally maximize the power in the ±1 diffraction orders (~35%–40%) and minimize the power in the zeroth diffraction order (<3%). For the last, this can be simply done by adjusting the etched depth in the phase mask production process.

When looking for advantages of the phase mask method compared with other methods, we may find that the grating period is independent of the UV wavelength; therefore, different UV sources can be used. Additionally, the grating period is independent of the exposure angle of incidence, requiring less accuracy in the alignment of the UV beam. Therefore, these features are ideal for mass production.

As for disadvantages, the defects presented in the phase mask are reproduced in the grating structure, which degrades the Bragg grating spectral response. Additionally, the inscription of different Bragg wavelengths requires the use of different phase masks. However, it is still possible to tune the Bragg wavelength a few nanometers by controlling the pre-tension of the fiber before the inscription. Another disadvantage of the phase mask method is that the phase mask needs to be placed very close to the fiber, this can lead to damage of the phase mask if improper mounting is performed or if any contaminant (e.g., dust) is present between the fiber and phase mask. To suppress this, a Talbot interferometer can be used in such a way that the phase mask is used as a beam splitter to create two separate beams which are then reflected by two mirrors and recombined to interfere at the fiber. If the mirrors are positioned at 90° angle, the grating produced in the fiber will be similar to that produced if the fiber was close to the mask. On the other hand, if the mirrors are rotated, other grating periods can be achieved [36].

2.5 FIBER BRAGG GRATING FABRICATION IN STEP-INDEX POF

The first photosensitivity study done on POFs dates back to 1998 [39] when Peng and co-workers illuminated a PMMA-based POF at different wavelengths. The fiber had been prepared through the "Teflon Technique" [40] and it was composed of two cores with 3 μm diameter each and separated by 10 μm distance, being the refractive index contrast manipulated by adding benzyl methacrylate (BzMA) to the core and trifluoroethyl methacrylate (TFMA) to the cladding, being fluorescein (170 ppm) added to the core mixture for the photosensitivity enhancement at the visible region. The role of each monomer can be classified as MMA to act as the main monomer; BzMA to increase the core refractive index and enhance the UV absorption due to the phenyl group; and TFMA, a low index fluorinated monomer used to reduce the cladding refractive index. Photosensitivity was observed through two methods, either axially through an Ar-laser operating at 633 nm or transversely with wavelengths set to 325, 280, and 248 nm from a frequency-doubled optical parametric oscillator (OPO) source, pumped by a frequency-tripled Nd:YAG pulsed laser. Regarding the former, the dopant gave rise to a large index change in the visible wavelength range. Concerning the latter, the index change was associated with the photosensitivity of the basic host material (PMMA) [39]. Furthermore, it was observed that the photosensitivity increased for shorter wavelengths (i.e., 248 nm).

Subsequent studies revealed the capability to photo-imprint refractive index changes in the core of polymer optical fibers, either MM [3] or SM [4]. Regarding

MM-POF [3], the authors produced a poly(MMA-co-EMA-co-BzMA), through the "Teflon Technique" and then, irradiated the fiber preform (the fiber core was uncovered by a prior polishing procedure), with lasers at wavelengths of 248 and 325 nm. The irradiation was done using a modified Sagnac interferometer that was constructed in order to get the interference of the two first-order diffraction beams of the source pump while blocking the zeroth-order. The irradiation was made through a phase mask with a period of 1061.4 nm and a laser output power of 5 mJ, with a pulse duration of 10 nanoseconds and repetition rate of 10 Hz. After the irradiation, two different results were obtained for each operational wavelength, namely, a surface relief grating for the 248 nm and a bulk grating within the core fiber for the 325 nm wavelength. The result attained for the 248 nm wavelength was associated with the high absorption of PMMA at that region, which resulted in a periodic removal of the polymer material at the surface. Because of the successful results obtained for the 325 nm wavelength, the authors decided to heat-draw the preform to fiber (core diameter ~30 μm) and then wrote an FBG using the same inscription setup. The result was a Bragg grating seen in reflection with a resonance wavelength centered at 1,570 nm and with a multi-peak behavior associated with the MM nature of the POF. The successful result obtained for the 325 nm wavelength drove the scientific community to use this wavelength region and exclude the 248 nm wavelength for the photo inscription of Bragg gratings in POFs [4,6,41,42].

Regarding the work reported in SM [4], the authors were able to report an FBG in a SM-POF composed of the same fiber materials and with core and cladding dimensions of 6 and 100 μm, respectively, [4]. The grating had 10 mm length, achieving a refractive index modification of 10^{-4}, with 80% reflectivity, 0.5 nm linewidth, and resonance wavelength at 1,570 nm [4]. Nevertheless, a few years later, the same authors were able to report grating reflectivity's as high as 99.8%, by an adaptation of the inscription setup and using a similar POF [41].

2.5.1 FIBER BRAGG GRATING FABRICATION WITH 325 NM HECD LASER

For the inscription of POFBGs using the most common wavelength (325 nm), a CW 30 mW HeCd laser, is commonly used. The inscription setup is normally based on the phase mask method, which can be assembled as shown in Figure 2.6.

As may be seen in Figure 2.6, the UV beam exiting from the HeCd laser ("A") is reflected by three mirrors ("B_n"). The first reflection occurs at the mirror ("B_1"), which is used to guide the beam upwards from the optical table. The second mirror ("B_2") is used to guide the beam parallel to the optical table, and the third one, ("B_3") is used to guide the beam downwards to the POF ("F"). Between the third mirror and the POF, there is a plano-convex cylindrical lens ("C"), that is used to focus the laser beam onto the core POF, increasing the power density in that region. The fiber is supported by a v-groove plate, placed on top of a 3D mechanical positioner ("H"). The latter is used to make a temporary free space coupling ("I") between the cleaved/polished POF and the silica pigtail fiber ("K"), terminated with an FC/APC contact, required to prevent Fresnel reflections. Between the fibers, a drop of an index matching gel is normally introduced in order to lower the background noise. The fibers need to be close enough to prevent the formation of FP cavities that can deteriorate

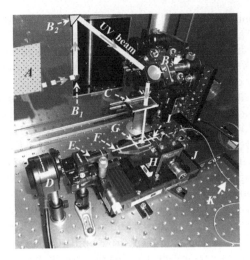

FIGURE 2.6 Picture of an inscription setup, using a CW HeCd, 325 nm UV laser. Legend: "A" → HeCd laser (not shown, behind the protective UV box); "B_n" → mirrors; "C" → lens; "D" → laser beam profiler; "E" → objective; "F" → POF (drawn with a white line); "G" → phase mask; "H" → 3D axis; "I" → butt coupling; "J" → motorised linear stage; "K" → silica pigtail fiber.

the visibility of the grating. To help the alignment process and to make sure that the light is propagated into the core of the POF, an objective ("E") is normally placed at the far end of the POF and the near field image can be observed using a laser beam profiler ("D"). The pattern imprinted into the fiber is determined by the period of the phase mask ("G") placed in close contact with the POF. The operation wavelength of the optical components shall suit the wavelength of the UV source, allowing to have high transmittance from the output laser to the fiber. The use of phase masks with operation wavelength other than the one of the laser source increases the intensity of the zeroth order, which is detrimental for the formation of the grating structure [43]. In such cases, the use of a modified Sagnac interferometer could be used, in order to block the zeroth-order and use only the ±1 orders [4].

The laser beam output of the HeCd laser has a circular shape with a diameter of 1.2 mm and typical output power of 30 mW [6], 50 mW [44], and 90 mW [45]. However, when the beam arrives onto the fiber, the power is slightly decreased due to the attenuation along the beam path. Furthermore, the beam also diverges, arriving with a small diameter difference compared to the laser beam exit. Considering a static irradiation system, the length of the POFBG is determined by the diameter of the laser beam at the fiber. Nevertheless, it is a common practice to use a lens (other than the one shown in Figure 2.6, marked as "C") to expand the UV beam along the length of the POF [6]. However, such an arrangement leads to longer inscription times as reported in Ref. [15]. Furthermore, it is also possible to inscribe gratings with longer lengths through the use of a motorized linear stage ("J" in Figure 2.6), allowing to move the 3D mechanical positioner, v-groove and fiber, all simultaneously, as reported in Ref. [14]. Another configuration to inscribe longer gratings is through the scanning of the UV beam along the length of the fiber [45,46], by moving, for instance, mirror B_1 in Figure 2.6, parallel to the length of the POF. Nonetheless, in both configurations (laser or fiber motion), it is necessary to perform the correct estimation of the velocity, allowing enough accumulated energy for each incremental motion which is necessary for an efficient refractive index modification in the fiber core.

The operational wavelength of the Bragg grating written in the POF ("G"), can be performed by proper choice of the period of the phase mask ("G") [14]. Despite the opportunity in inscribing gratings at the low loss region of polymers (~600 nm), a vast amount of the works reporting POFBG has been made for the C+L regions, where most of the light sources and detectors are available at a lower cost, due to the already developed fiber telecom industry. Working with POFs at this wavelength regions (excluding perfluorinated fibers) is challenging since the attenuation is too high and thus, POF lengths are limited to less than 10 cm.

2.5.2 FIBER BRAGG GRATING MEASUREMENT SCHEMES

To monitor the grating growth, it is possible to use two configurations, either in transmission or in reflection, as shown in Figure 2.7a and b.

Concerning the measurement scheme shown in Figure 2.7a, specifically for the transmission scheme, a broadband source is used to inject the light into the core POF and the signal output is collected by an optical spectrum analyzer (OSA), at the rightmost part of the figure. On the other hand, for measuring the FBG reflection signal, the broadband light source is injected in one arm of a three-port fiber coupler/circulator. The output signal is then coupled to the POF core, and the FBG reflection spectrum is observed by connecting the third arm of the fiber coupler/circulator into the OSA. Since this scheme requires only one single silica to polymer optical fiber splice, it is, therefore, the most adopted. Furthermore, the reflection method is more sensitive, since reflections right above the noise floor of the OSA are easily displayed [47].

Nevertheless, commercially available interrogators can also be used to monitor the Bragg reflection signal in a more affordable way, since the device has an already built-in source, circulator, and detector (see Figure 2.7b). Thus, the FBG reflection

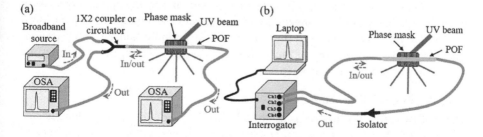

FIGURE 2.7 Schemes used to measure the POFBG reflection and transmission signals, using a laser source, coupler/circulator and an optical spectrum analyzer (a), or through the use of a fiber interrogator (b).

signal can be easily measured by simply plugging the fiber terminal on the input/
output (In/Out) channel of the interrogator. Moreover, if the interrogator contains
more than one In/Out channel, then, the transmission signal can also be measured.
This can be easily accomplished by using one of the channels as a source and the
other, just as detector. For that, a fiber isolator can be used at the output of one of the
interrogator channels. In this way, one can simultaneously measure the reflection and
transmission signal in each of the interrogator channels, as represented in the scheme
presented in Figure 2.7b.

2.5.3 GRATING GROWTH BEHAVIOR

In literature, fiber Bragg gratings written in silica optical fibers can be classified
according to their growth mechanism. The most basic types of grating associated
with UV inscription are the type I and type II. Regarding type I gratings, they are
referred to the ones where the refractive index modulation occurs below the damage threshold; while for type II, the refractive index changes occur above the damage
threshold. These two types of gratings can also be grouped in a variety of sub-
classes that are identified according to the writing conditions and properties of the
gratings [36]. Even though, the mechanisms behind refractive index change can be
very different [17].

Type I gratings are the most common gratings found in literature, and they have
the particularity of showing a reflection spectrum with a negligible light loss to the
cladding modes or by absorption. The index modulation attained in this type of gratings is lower than 10^{-4}, due to the low energy (E) deposited by each laser pulse. On
the other hand, above a certain exposure threshold, the index modulation increases
drastically and becomes saturated. This type of grating is referred as type II and was
first demonstrated by Archambault [48], who showed the possibility to write Bragg
gratings with reflectance higher than 99.8%, with a single high energy pulse. The
index modulation for such gratings can achieve values as high as 10^{-2} [48]. Despite
the opportunities of implementing an inline grating fabrication process as the fiber
is drawn from a preform, which promotes the fabrication of low-cost devices due
to the single shoot nature, these gratings have not been receiving so much attention
in the telecommunication domain due to the poor transmission spectrum, which
comprises high bandwidth and light loss at the blue region of the Bragg wavelength
[36,48]. On the other hand, they have been important in sensing applications due to
the higher stability at high temperatures [36,48]. Additionally, microscopic exami-
nation of type II gratings show the presence of a periodic damage track (micro-
fractures) within the vicinity of the grating and for that reason, these gratings are
also known as damage gratings [36,48].

2.5.3.1 POFBG Growth Behavior

The POFBG growth behavior using the 325 nm UV radiation is well documented
and is generally known to be a time-dependent process that takes several tens of
minutes to achieve grating saturation [17,18,41,49]. The reason for such a long time is
mainly due to the low photosensitivity of PMMA at 325 nm. The laser power and the
repetition rate in the case of pulsed lasers are also important for the proper writing

of POFBGs [18,50]. Nevertheless, the scattering of the UV beam at the air holes in the case of microstructured fibers needs also to be taken into account.

Generally, when using 325 nm UV lasers for the POFBG inscription, the visualization of the first reflection peak only occurs after a few minutes. Then, the peak power reflectivity increases up to the saturation. Concerning to the resonance Bragg wavelength, it is commonly observed a blue wavelength shift during the irradiation process. Right after the inscription, a small red wavelength shift is also described, being its stabilization reached after a few minutes, where the final Bragg wavelength is lower than the one when the first Bragg peak appeared. The reflection spectra, reflection peak power, and Bragg wavelength obtained in a typical POFBG inscription process, using the 325 nm CW HeCd laser with 30 mW output power and through the phase mask technique, ($\Lambda_{pm} = 1048.7$ nm), during and after the inscription, may be seen in Figure 2.8a–c, respectively.

As may be seen from Figure 2.8b, in the beginning, the reflection peak power is unnoticed. Conversely, after ~7 minutes, the Bragg peak starts to appear, showing a resonance Bragg wavelength located at ~1,539 nm (see Figure 2.8c). Overall, at the beginning, the grating grows with a well-defined peak, as a result of the induced refractive index change in the fiber core only, being thus, classified as a type I grating according to Liu et al. [17], who follow the descriptions already reported for the grating formation in silica fibers. However, as the grating grows, the spectra bandwidth becomes broader and the losses at the short-wavelength increase (Figure 2.8a). This specific behavior is described as type II grating [17]. In our example, after 30 minutes the Bragg peak reached saturation and the laser was turned off, as seen in Figure 2.8b. The peak power drops in the next few minutes and then starts to increase again until stabilization at the similar reflectance as when the irradiation was off. In literature, it may be found that continuous exposure of the grating after saturation leads to a decrease of the reflectivity, which becomes negligible in long exposure (~30 minutes) [18]. However, grating regrowth has been visualized after the laser is switched off, reaching its saturation after a few hours. In an attempt to explain the results, the authors associate the phenomenon to the thermal stress created during the irradiation process that could induce a refractive index change with an opposite sign to that induced by the UV photoreactions. Since the refractive index change due to photoreactions is permanent and the thermal one could be relaxed over time, then, regrowth of the grating could occur [18].

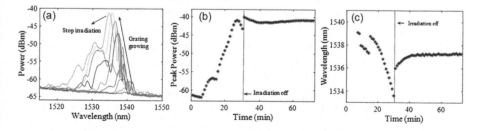

FIGURE 2.8 Spectra of the Bragg grating growth during the POFBG fabrication (a), reflection peak power (b), and wavelength (c) during and after the inscription process.

From Figure 2.8c, one can see the Bragg peak wavelength over time, which demonstrates the net contribution of the irreversible and reversible processes, associated with the permanent refractive index change due to the UV irradiation and to the thermal change due to the rise of temperature on the irradiated region, respectively. The irreversible change can only be accounted after the fiber is cooled down, which is indicated by the stabilization of the Bragg wavelength, 10 minutes after the irradiation is stopped. The remaining permanent and negative Bragg wavelength shift indicates a negative mean refractive index change, which is the result of the PMMA main chain scission, as concluded in Refs. [24,25] for the 320 nm wavelength radiation. Therefore, shorter molecule chains are produced, leading to a higher degree of freedom and consequently, lower density, which in turns, results in a decrease of the refractive index.

Temporal stability of Bragg gratings is a key factor for the construction of any type of device. Harbach [22] observed that the grating reflectivity in POFBGs written with a 308 nm Xenon monochloride (XeCl) excimer laser, became stable only after a 6-month period, assuming that the induced refractive index change took several months to reach stability. Later, Statkiewicz-Barabach et al. [9] decided to inscribe type I and II gratings in POFs using a 325 nm CW HeCd laser and verify their long-term stability during an 8-month period [9]. Their results have shown that type I gratings decrease their reflectivity in the next hours or days after the inscription and, in some cases, disappeared. On the other hand, type II gratings showed only a slight change of the reflectivity over time.

The stability of type I and type II gratings inscribed in silica fibers are known to be temperature-dependent, being the type II gratings more suited to reach higher temperatures. However, Statkiewicz-Barabach et al. observed that PMMA FBGs of either type degrade at temperatures of ~85°C and disappear completely at 90°C [9].

2.5.3.2   Fiber Tapering Influence on the POFBG Inscription

The grating reflectivity and inscription time do not depend exclusively on the photosensitivity of the material and wavelength source. In fact, the amount of polymer material surrounding the core attenuates an important part of the UV radiation, which means that the diameter of the POF has a strong impact on the efficiency of the grating inscription. Rajan et al. first described the inscription of POFBGs in tapered SI-POFs, showing that the inscription time was reduced after reducing the diameter of the fiber to approximately 16 μm [44]. The fiber used in their work had a PMMA cladding diameter of 240–260 μm and a 12 μm core, made of poly (MMA-co-EMA-co-BzMA. The diameter reduction was first made by immersing the POF in a mixture of acetone and methanol for 14 minutes, allowing to achieve a fiber diameter of 27 μm. Then, the etched fiber was tapered through a heating stage to achieve a final diameter of 16 μm. The Bragg grating was imprinted through the phase mask technique using a 50 mW, 325 nm CW HeCd laser, allowing to obtain a POFBG at the 1,550 nm region in ~3 minutes which was lower than the 7–10 minutes normally required to imprint a POFBG in an untapered POF [44]. Despite the reduction on the inscription time, the grating reflectivity was about 5% which was lower than the one normally achieved for untapered fibers ~50% [51]. The reduction on the inscription time was according to the authors, dependent on the amount of photosensitive core region left in the fiber after the tapering process, together with the surface modification during the etching and tapering process [44]. Nevertheless, the amount of PMMA material removed during the chemical etching allows the UV beam to reach the core POF with a higher power, allowing to reach the required UV dosage in a lower period of time. Hu et al. [46], in an attempt to increase the POFBG grating reflectivity and inspired by the work developed by Rajan et al., decided to slightly etch a photosensitive POF. The fiber had ~150 μm cladding diameter and was composed of undoped PMMA, while the core had 8.2 μm diameter and was composed of PMMA doped with trans-4-stilbenemethanol (TSB) (1% wt) and diphenyl sulfide (DPS) (5% mole). Both dopants were used to increase the refractive index and enhance the photosensitivity. Furthermore, DPS facilitates the trans–cis interconversion under UV irradiation [52]. The fiber was slightly etched in a mixture of acetone and methanol (2:1) to reach a diameter of 133 μm [46]. The inscription through the scanning phase mask technique was then followed, using a 30 mW, 325 nm CW HeCd laser, a total exposure time of 40 minutes and a grating length of 6 mm, giving a scanning speed of 2.5 μm/s. The result was a POFBG centered at the 1,550 nm region with reflectivity as high as 97%, and refractive index modulation equal to 2.5×10^{-4}, which was much higher than the 25% reflectivity achieved for untapered POF, using the same inscription procedure [46].

Later, Bhowmik et al. [53] replicated the results performed by Hu et al, showing a reflectivity increase and inscription time decrease with reduction in the fiber diameter from 185 μm, down to 85 μm, achieving a higher reflectivity of 98.5%, in about 7 seconds of exposure time, for a 10 mm grating length written through the phase mask technique and using a 50 mW 325 nm CW, HeCd laser.

2.5.4 BRAGG GRATINGS IN PHOTOSENSITIVE CORE DOPED SI-POFs

Photosensitivity in POFs has been firstly described for the inscription of a FBG in a PMMA based optical fiber with copolymer core of MMA-EMA-BzMA [4]. The photosensitivity for other undoped POF formulations has also been reported, such as PMMA fibers with a core composed of PMMA/PS copolymer [9,22,49,54,55]. Figure 2.9d and e shows the reflection spectra of two gratings written in these type of fibers at the 850 and 1,550 nm regions, respectively, using a 325 nm CW HeCd laser. The fiber was a SI-POF from Paradigm Optics, Inc. with trade name MORPOF03, with a core made of the copolymer PMMA/PS with less than 5% PS and with core and cladding diameters of 8 and 125 μm, respectively. The microscope image, as well as the simulated fundamental mode propagation and near field pattern of the fiber, are shown in Figure 2.9a–c, respectively.

The use of the conventional 325 nm CW HeCd laser source for the inscription of POFBGs with good reflectivity normally requires long exposure periods [17,41]. However, the long exposure causes material damage due to photodegradation. The process is not economic, requiring stabilized systems during the inscription process and the resulting grating can be physically and mechanically weakened [9,17], which can compromise the integrity of the final application.

As a result of the low photosensitivity of PMMA material under 325 nm wavelength, functional dopants, such as fluorescein [3], trans-4-stilbenemethanol

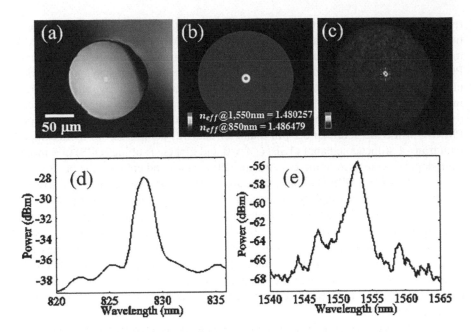

FIGURE 2.9 POF microscope image (a), simulated mode field propagation (b), and near field pattern of an undoped PMMA SI-POF with core copolymerized with 5% PS. The reflection spectra of a POFBG written in this type of fiber at 850 and 1,550 nm regions is shown in (d) and (e), respectively.

(TSB) [42], benzildimethylketal (BDK) [56], methyl vinyl ketone (MVK) [57], 9-vinylanthracene (9-VA) [58], benzophenone (BP) [59], DPS [52], and recently diphenyl disulphide (DPDS) [60], have been introduced in the core POF for the photosensitivity enhancement. The incorporation of these dopants in the core fiber has been made by co-polymerization with photosensitive monomers [57,58,61,62] or through simple doping with small molecules [3,42,56].

Photosensitivity can be defined as reversible and irreversible. Examples of the former may be found for azobenzene [61,63], in which gratings can be formed by photo-induced birefringence. For that, the azobenzene units are aligned through linearly polarized light and erased through circularly polarized light, creating and erasing the birefringence and hence fiber gratings. Photosol® 7–049 is another dopant example of reversible photosensitivity [64]. In this case, the photosensitive molecules can alternate between two isomeric forms with a different refractive index. The transition is induced by light towards the "excited" state and by visible light or heat back to the relaxed state. Moreover, this process is totally reversible and, therefore, Bragg gratings can be created with possibilities to be switched on and off [64]. Despite the opportunities presented for the reversible photosensitive materials, most of the work has been developed for the irreversible ones. In this case, the photosensitivity could be linked to photolysis [30], photolocking [65], photo-polymerization [56,64], photocrosslinking [64], and also photo modification through femtosecond lasers [31,32]. An extensive report of the photosensitive mechanisms reported so far may be found in Ref. [66]. Moreover, studies on the improvement on the photosensitivity have also been reported for etched fibers, which can be attributed to the less amount of PMMA material that attenuates the UV radiation before reaching the core POF [46]. Another important factor that is well known for increasing the photosensitivity in PMMA material is the tensile and shear stress present in the polymer material [67]. Studies confirming this property were done by Sáez-Rodríguez et al. [68], who assumed that the increase in photosensitivity of PMMA fibers kept under stress and irradiated with a 325 nm UV radiation is due to a photodegradation process that generates monomers and free radicals that are later used as initiators in the polymerization process.

2.5.4.1 Gratings in TSB Doped POFs

Polymers doped with isomers can present different refractive index, depending on the state of the isomer, which in turn, depends on the UV radiation. According to Ref. [42], the isomer TSB was used as a dopant in the core of a PMMA fiber. TSB is a stilbene derivative, which can undergo reversible trans-cis photoisomerization. The photochemical transformation process of the 4-stilbenomethanol from a trans- to cis-structure upon UV radiation can be seen in Figure 2.10. The refractive index of these compounds is different, where the trans-species present a higher refractive index value.

The photoisomerization was first observed for films doped with TSB illuminated with a 320 nm radiation from a high-pressure mercury lamp. Results revealed a refractive index decrease, which was accounted to be between -3×10^{-4} and -9×10^{-4} for a dopant concentration of 1 and 2 wt% [42]. Despite small, the refractive index modification was enough to create a Bragg grating with a strong reflection signal. For that, the authors used a POF system comprising a cladding made of a copolymer of MMA and BMA (90/10 molar ratio) and a core made of a TSB-doped MMA–EMA–BzMA (90/4/6 molar ratio + 0.66 wt% TSB) copolymer, being the diameters of the core and cladding regions of 20 and 250 μm, respectively and the refractive index difference of 9×10^{-3}. The grating was written through the UV scanning phase mask technique, with a scanning rate of 0.12 mm/minute, through a 325 nm laser operating at 10 Hz repetition rate, with 10 nanoseconds pulse width and fluence of 20.8 mJ/cm², together with a phase mask period of 1046.5 nm. The result was a POFBG centered at 1,550 nm region, with a length of 11 mm, during an accumulated exposure of 10 minutes. The grating was seen in transmission as shown in Figure 2.11, showing several dips at the 1,550 nm region due to the several modes

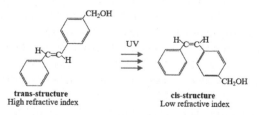

FIGURE 2.10 4-stilbenemethanol isomerization from trans- to cis-structure upon UV irradiation.

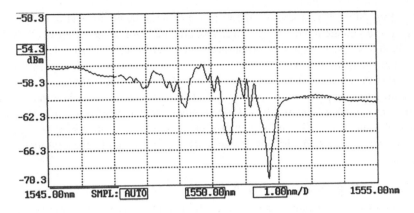

FIGURE 2.11 Transmission spectrum of the first POFBG recorded in a SI-POF with core doped with TSB. (Reprinted from J. Yu, et al., "Trans-4-stilbenemethanol-doped photosensitive polymer fibers and gratings," *Opt. Lett.*, vol. 29, no. 2, pp. 156–158, 2004, with permission from the Optical Society of America.)

allowed to propagate in the fiber. The transmission dips had strength up to 9 dB, corresponding to a reflectivity of up to 87% [42].

Attempts made by the same group, to imprint FBGs using undoped POF (free of TSB), made of PMMA cladding and core of MMA–EMA–BzMA (88/4/8 molar ratio), with core and cladding diameters of 9 and 140 μm, respectively, revealed that even reducing the scanning rate to 0.03 mm/minute and increasing the fluence to 29.2 mJ/cm^2, the grating started to appear only after several hours, reaching its saturated state after ~32 hours. Thus, the effectiveness of the TSB on the photo-imprinting process of FBGs was confirmed [42,69]. Furthermore, the ability to imprint LPGs in TSB doped POFs through the amplitude mask technique with a period of 0.35 mm, using the same laser, and with a fluence of 121 mW/cm^2, during 15 minutes of exposure time, was later reported by the same authors [70].

Due to the high attenuation of PMMA at the 1,550 nm region, Zhang at al. [45] decided to imprint a POFBG at shorter wavelengths, namely at 962 nm. The work was made using a TSB doped POF in which the core and cladding had 8 and 130 μm, respectively. The irradiation was made using a CW 325 nm HeCd laser with an output power of 90 mW. The grating was imprinted through the phase mask technique, with a phase mask period of 647 nm (suited for the 248 nm wavelength), and with a scanning velocity of 4 μm/second, taking about 15 minutes to reach a Bragg wavelength at 962 nm with few peaks in reflection due to the few-mode (FM) nature of the fiber. A reflectivity of about 20% and a refractive index modulation amplitude of 1.5×10^{-5} was obtained [45]. The lower reflectivity when compared with the 87% reported in [42] can, in fact, be related to the operational wavelength of the phase mask used in Ref. [45], which was different from the one of the laser source (i.e., $248 \neq 325$ nm). This led to non-negligible zeroth diffraction order, which reduces the visibility of the grating pattern imprinted in the core of the POF.

The stability of POFBG over time is critical and its study for undoped POFs under different exposure conditions has been described in Refs. [9,18,22,71]. Based on that, Hu et al. [72], decided to study the influence of the writing power on the long-term stability of TSB doped POFBGs. In their study, they observe that for gratings written under high power densities, the grating reflectivity was stable over time. Conversely, for low power densities, the grating peak power decreased over time. However, it was shown that using a post-annealing treatment it was possible to recover and to keep stable the grating reflectivity [72]. Furthermore, using the same thermal treatment, it was possible to show gratings with 25 dB reflection band above the noise level, with just 1 second of inscription time. The explanation of the results was given through the transition mechanisms between the excited intermediate states of both trans- and cis-isomers and the temperature-dependency glassy polymer matrix explained in Ref. [73].

2.5.4.2 Gratings in BDK Doped POFs

Another well-known photoactive dopant commonly described in the literature for efficient POFBG inscription is BDK. This dopant is a classic photosensitive dye and when irradiated with UV radiation, there is enough photon energy to activate the n–π* transition within the molecular orbits of the >C=O group, that is then followed by an α-splitting to produce free radicals [65,74] as shown in Figure 2.12a. The radicals can then react with polymerizable methacrylic groups as shown in Figure 2.12b, creating shorter PMMA molecules and increasing the refractive index. Furthermore, the radicals produced by the photodecomposition of BDK can undergo coupling of two benzoyl radicals to make a benzyl molecule as shown in Figure 2.12c. The molecule is also photosensitive and further photochemical reaction during UV illumination can occur. The final products formed in the illuminated area, in an oxygen-containing atmosphere, do not evaporate or crystallize, being locked or fixed in the polymer matrix, a phenomenon called "photolocking" [74]. Thus, since the radicals produced by the photodecomposition of BDK have high electronic polarizability, due to the phenyl group, they can increase the refractive index in the irradiated areas [65,74].

Comparisons of the UV absorption of BDK with that of undoped PMMA revealed new absorption bands for the former, which were centered at 250 and 344 nm [56], attributed to the π–π* and n–π* transitions, respectively.

In an attempt to develop a highly photosensitive POF for efficient Bragg grating inscription, the group of Peng et al. [56] decided to investigate the photosensitivity properties of PMMA and BDK doped PMMA samples when continuously exposed to 365 nm wavelength source. The results revealed a continuous reduction on the peak absorption located at 250 nm, as a result of the photodecomposition process of BDK. On the other hand, undoped PMMA samples showed no significant spectral changes for the same exposure conditions. The results were indicative of the BDK photosensitivity at 365 nm wavelength. Because of that, the authors decided to fabricate two SI-POFs, either SM or MM, through the "Teflon technique," obtaining the fiber specifications described in Table 2.1.

The recording system was based on a modified Sagnac optical ring system using a 355 nm frequency-tripled Nd:YAG pulse laser with 6 nanoseconds pulse width and operating at 10 Hz. The gratings were imprinted through a phase mask with period of 1,061 nm and with a beam spot of 6 mm. The resultant grating for the MM fiber was

(a) Photodecomposition of BDK

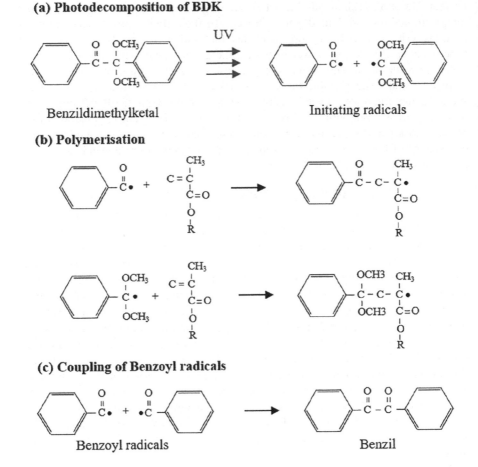

Benzildimethylketal   Initiating radicals

(b) Polymerisation

(c) Coupling of Benzoyl radicals

Benzoyl radicals   Benzil

FIGURE 2.12   Possible photochemical reactions of BDK.

TABLE 2.1
Fiber Specification Used in Ref. [56]

Type	Core Material	Cladding Material	wt% BDK	Core Diameter	Cladding Diameter	Refractive Index Difference
MM	MMA-BA	MMA-BA	9.6	21.0	290.0	0.011
SM	MMA-EMA-BzMA	MMA-EMA	2.0	11.8	230.4	0.001

---

seen after 16 minutes, for a power density of 673 mW/cm^2. The POFBG was located at 1,570 nm and presented multiple peaks, reaching a reflectivity of about 25%. The corresponding refractive index modulation was about 4.5×10^{-5}, which was lower than the value reported for BDK doped PMMA samples, which was 2.4×10^{-3} [74].

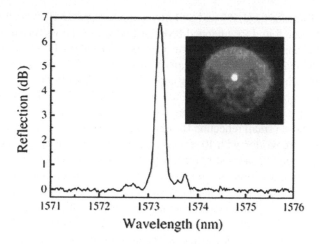

FIGURE 2.13 Spectrum of a POFBG in a SM-SI PMMA fiber with core doped with BDK, inscribed with a 355 nm Nd:YAG laser source. The inset shows the fiber used in the experiment. (Reprinted from Y. Luo, et al., "Gratings fabrication in benzildimethylketal doped photosensitive polymer optical fibers using 355 nm nanosecond pulsed laser," *Opt. Lett.*, vol. 35, no. 5, 751–753, 2010, with permission from the Optical Society of America.)

On the other hand, the grating on the SM-POF was written with low fluence (200 mW/cm^2) to avoid damages on the POF. The grating was written in 17 minutes and was seen in reflection as shown in Figure 2.13, having a single peak located at 1,570 nm and presenting 6 dB above the noise level [56].

Subsequent works based on FBGs in BDK doped POFs were reported by other authors in both SI [75,76] and microstructured POFs [77,78], using different laser sources such as 355 nm frequency-tripled Nd:YAG pulsed laser, 325 nm CW HeCd laser [76–78] and recently laser sources with wavelength close to the main absorption band of BDK (250 nm), namely: 248 nm KrF pulsed laser [79] and 266 nm of the 4th harmonic of a Nd:YAG pulsed laser [80]. Inscriptions with one laser pulse and with refractive index modulations as high as 0.7×10^{-4} and 0.5×10^{-4}, were achieved for the 248 and 266 nm laser sources, respectively.

2.5.4.3 Gratings in DPDS Doped POFs

Recently the fabrication of a PMMA SM-POF, in which the core was made of PMMA (92 wt%) doped with DPDS (8 wt%), has been reported for the fast Bragg grating inscription [60]. The dopant was chosen to increase the refractive index of the core and also to enhance the photosensitivity under UV exposure. The fiber was produced through the "pull-through" method, in which the core preforms are polymerized in a container and then, heated and drawn to tightly fit the hollow cladding preform previously prepared through the "Teflon technique." With this approach, the authors were able to avoid the dopant diffusion that commonly occurs in the "Teflon technique." This allowed to attain an abrupt refractive index change between the core and cladding regions, and thus, easier to achieve the single-mode behavior in SI-POFs. The final fiber had a core and cladding diameters of 5.5 and 120 μm, respectively, and an attenuation of 87 and 27 dB/m for the 1,550 and 850 nm regions, respectively, which was slightly higher than the one found for undoped PMMA fibers. For the grating inscription, the authors used the phase mask technique through the conventional CW HeCd laser operating at 325 nm and with a phase mask pitch of 1,046 nm, intended for gratings at the 1,550 nm region. The laser beam arriving at the fiber had ~26 mW of optical power, after passing through a beam expander used to cover an FBG length of 10 mm. The gratings were written with different exposure times, where a minimum of 7 milliseconds was needed to show a small reflection Bragg peak, right above the noise level. Stronger Bragg gratings were written with 10 seconds exposure time. The corresponding results may be seen in Figure 2.14a and b, for the gratings written right after the inscription process with an exposure time of 7 milliseconds and 10 seconds, respectively [60].

The result obtained in Ref. [60], and shown in Figure 2.14, established a milestone on the current inscription time need to achieve a POFBG with the 325 nm UV source, which was about 5–7 minutes for fibers of similar diameters [10,15,81] or 3 minutes for thin diameter fibers [82]. Nevertheless, the authors also observed that the gratings reflectivity changed after the writing process, reaching stabilization only after about 1 week. The phenomenon was explained by a possible molecular reorientation of the polymer chains [60] as also explained by other authors [78].

In an attempt to explain the photosensitive mechanisms involved in the grating formation, the authors considered the homolysis of the DPDS under UV radiation, with the formation of two sulphenyl radicals, as shown in Figure 2.15, which can then bond to the photodegraded sites on the PMMA chain (see Figure 2.1), through the Sulphur atom [60].

Theoretical predictions were later confirmed by Raman spectroscopic measurements, in which the Raman peaks associated with the S–S and C–S–S bonds decreased, while a slight increase was observed for the C–S bonds: breaking a DPDS molecule into two, can potentially generate twice as many C–S bonds if the sulphenyl radicals join onto a PMMA chain [60].

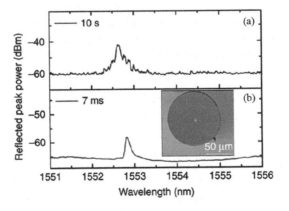

FIGURE 2.14 Reflection spectra obtained for DPDS doped SM SI-POF, right after the inscription process, for an exposure time of 7 milliseconds (a) and 10 seconds (b). The inset figure corresponds to the microscope image of the fiber. (Reprinted from J. Bonefacino, et al., "Ultra-fast polymer optical fibre Bragg grating inscription for medical devices," *Light Sci. Appl.*, vol. 7, no. 3, pp. 17161, 2018, with permission from Springer Nature.)

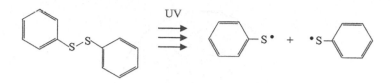

FIGURE 2.15 Homolytic fission of the DPDS with the formation of two sulphenyl radicals.

2.6 FIBER BRAGG GRATING FABRICATION IN MICROSTRUCTURED POLYMER OPTICAL FIBERS (mPOFs)

Microstructured optical fibers are a special type of fibers where the light is guided through special hole arrangements in the cross-sectional area of an optical fiber. The arrangement of these holes, out of the central area, allows having a depressed index at the hole region compared to the solid inner part. Therefore, the condition for total internal reflection, or equivalently mode confinement, is satisfied [83]. This type of fiber is commonly known as photonic crystal fibers and was first introduced in 1996 [84]. Both hole dimension (d) and hole to hole distance (Λ_d) can be adjusted to offer special characteristics such as, the ability to be SM in a wide range of wavelengths, also known as endlessly single-mode operation [85]. Furthermore, other arrangements can be used, either by designing asymmetric structures, in order to provide high birefringence [86–90]; or the control of the air filling fraction (d/Λ_d), in order to adjust the dispersion [91]. The fabrication of silica microstructured fibers is essentially realized by the stack-and-draw-technology, which is time-consuming and does not allow the creation of special hole arrangements. On the other hand, the practical implementation of this kind of structures in POFs is easy to create with flexible technologies, such as drilling [5], preform casting [92] and 3D printing [93]. However, fibers with SM behavior (that is usually preferred in FBG sensing applications due to their single peak nature, which offers higher resolution) are difficult to achieve with SI profiles. This occurs due to the compromise between the refractive index contrast and the core radius which needs to be balanced in order to have a *normalized frequency* lower than 2.405. Doping diffusion is the key problem especially when SM behavior in the low loss region of polymers is desired.

The development of the first mPOFs in 2001 [5] solved many of the problems associated with the SI profile allowing to explore different characteristics, such as gradedindex (GI), high-birefringence, dual core, suspended core, hollow core, etc. [94]. Furthermore, the possibility of doping those fibers can also be attained [95] allowing, for instance, to enhance the photosensitivity properties of those fibers [77,78,96].

2.6.1 INFLUENCE OF THE MICROSTRUCTURED AIR HOLES ON THE POFBG INSCRIPTION

Dobb et al. were pioneers in demonstrating the first FBGs in an mPOF [6]. The authors used a fiber with FM behavior composed of four layers of air holes disposed hexagonally and with the central hole missing (see a representation of the mPOF in Figure 2.16).

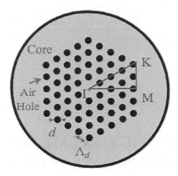

FIGURE 2.16   Representation of the 4-hole ring structure mPOF used in Ref. [6].

The fiber was drawn without sleeving the secondary preform, which was commonly used on the mPOF fabrication, providing a small diameter fiber and thus, lower absorption by the UV writing beam. The final mPOF had a diameter of 160 μm, with holes of 5.5 μm diameter and hole to hole distance of 10 μm, giving an air filling fraction of 0.55. The grating inscription followed the phase mask technique, using a 325 nm CW HeCd laser with 30 mW output power and a phase mask with a period of 1060.85 nm, allowing to get a resonance Bragg wavelength at the infrared region. The laser beam had 1.8 mm diameter and it was expanded in the longitudinal axis of the mPOF and focused on its transversal section by two consecutive lenses, in order to produce gratings in the core region with a total length of 10 mm. The grating inscription took 60 minutes to achieve the maximum reflectivity. The obtained POFBG was centered at 1,570 nm, presenting a multi-peak behavior, a bandwidth of 1 nm and a reflective power above the noise level of 7 dB. A similar procedure was also conducted for a SM mPOF presenting $d/\Lambda_d = 0.31$, allowing to get a POFBG centered at 1,569 nm, with a bandwidth of 0.5 nm and a reflective power above the noise level of 6 dB [6]. The authors also report the importance on the orientation of the microstructure with respect to the incident beam, revealing that the successful inscription was that in which the fiber orientation produced a thin and bright back reflection when irradiated. The conclusion was associated with the scattering of light at the cladding air holes region, which was assumed to be lower at the angular orientation labeled ΓM (see Figure 2.16).

The use of fibers with flat sides at the orientation perpendicular to the ΓM direction were also developed by the same group with an aim to aid the alignment of the laser beam relatively to the microstructured cladding [97,98] (see the fiber representation shown in Figure 2.18a). However, the authors did not observe any benefit. Another idea to decrease the amount of scattering at the microstructured region was based on the use of only two layers of air holes in an hexagonal arrangement [99]. The fiber, which is represented in Figure 2.18b, was composed of TOPAS® material and had a cladding diameter of 240 μm, a hole diameter of 2 μm and a pitch of 6 μm, providing a d/Λ_d of 0.33. Despite the interesting idea, the reported exposure time was over 5 hours. Possible reasons for such result may be attributed to the low photosensitivity of the polymer material under UV radiation. However, it is mostly believed that the major cause is associated with the lower intensity of the writing beam in the core region, which is associated to the scattering of the UV radiation on the hole layers that are still present in the fiber, as it is reported by different works [49,54,100].

One obvious consequence of a long exposure process, on the POFBG inscription, is the necessary stability during the whole inscription process. However, in 2014, Bundalo et al. reported that the conventional inscription technology (phase mask technique through a 325 nm UV laser), could achieve a record inscription time of less than 7 minutes [15]. According to the authors, the success was due to the higher power reaching the core due to a modification on the laser inscription setup associated with the absence of a secondary lens used to expand the UV radiation along the length of the fiber. To get evidences about the influence of the laser power on the recording time, the authors placed an attenuator in the path of the UV beam, allowing to write gratings with different powers. The results showed a fast inscription time for higher powers, reaching values close to 7 minutes [15]. Despite the record inscription time, the authors did not take into account the orientation of the UV beam against the microstructured air holes. At that time, previous studies developed by Marshal et al. [101] in silica PCFs have shown both numerically and experimentally that the scattering effect of the air holes structure was minimum for the ΓK orientation. The conclusions were taken from the measurement of the red luminescence at the fiber output, created in the core region when the fiber was exposed to UV light. Thus, it was possible to observe higher photoluminescence when the direction of the UV radiation was close to the fiber ΓK direction. Oliveira et al. reported that the transverse coupling of UV radiation in the fiber core of an mPOF could also be observed through the visualization of the near field image of the tip of an mPOF during the POFBG inscription process [102]. For that, a microscope objective was placed at the end of an mPOF and a laser beam profiler collected the UV radiation scattered in the fiber, as can be seen in Figure 2.17a and b, respectively.

One year after the demonstration of the possibility to write FBGs in mPOFs in less than 7 minutes by Bundalo et al. [15], the same group decided to study the influence of the FBG fabrication at different mPOF orientations [103], allowing to corroborate their results with the work developed by Marshal et al. For that, the POFBGs were written up to the saturation (>15 minutes) and then, the exposure

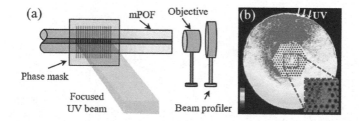

FIGURE 2.17 (a) Setup used to collect the near field image, composed of an objective and a laser beam profiler placed at the end of an mPOF under UV exposure. (b) Near field of the mPOF under the UV exposure. The arrows indicate the UV direction, being the darkened color associated with the UV radiation passing through the fiber that can reach the core as seen by the inset. (Adapted from R. Oliveira, L. Bilro, J. Heidarialamdarloo, and R. Nogueira, "Fast inscription of Bragg grating arrays in undoped PMMA mPOF," in *24th International Conference on Optical Fiber Sensors*, 2015, vol. 9634, p. 96344X.)

continued up to 30 minutes. The long exposure allowed to burn the POF region on the direction of the UV radiation, which was then revealed by cleaving the fiber perpendicularly to its length. Results showed that FBGs in mPOFs could be imprinted at any angle related to the mPOF microstructure, contrary to the observations taken by Dobb et al. [6], for the first report of an FBG in an mPOF. Nevertheless, they observed that the angles close to the ΓK orientation were the ones that showed better reflectivity and faster recording times, which was in accordance with the results predicted by Marshal et al.

One interesting advantage of microstructured fibers is their ability to tailor specific applications by proper design of the air-hole structure. Therefore, the possibility of exposing the fiber core region to the writing beam, by manipulating the number of air holes and/or the air filling fraction, is interesting and was readily proposed by two different groups, for an undoped PMMA mPOF, in which part of the air holes were omitted [19], (see Figure 2.18c); and for a 3-hole photosensitive mPOF [104], (see Figure 4.8d).

Regarding the work developed in Ref. [19], the POFBG was inscribed with the point-by-point technique through a femtosecond laser operating at 800 nm. The work was accomplished with the use of an exotic mPOF composed of 3-rings of air holes, in which some of the holes were omitted for efficient UV access to the core region, (see Figure 2.18c). Such fiber was made by removing two air holes in the outer ring and one in the second ring, being a total of six holes removed due to the symmetry of the fiber, allowing the access to the focused writing beam on both sides. Furthermore, the distance between the air holes and the diameter of the mPOF was adjusted to allow the focused beam (~1 μm), to go through the bridge, as well as to allow SM operation at 1,550 nm region. The grating was written by aligning the "opening" air hole region to the femtosecond laser beam. This step was facilitated through the visualization of an intentional groove at the outer region of the fiber, as can be seen in Figure 2.18c. The laser repetition rate was 1 kHz and a 4th order grating was chosen to avoid the overlap of two consecutive spots. The grating was fabricated at the 1,520 nm by using a fiber translation speed related to the laser beam of 2.053 mm/s, allowing to inscribe a 5 mm length POFBG in just 2.5 seconds [19].

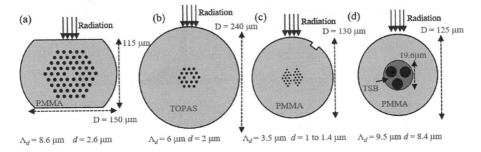

FIGURE 2.18 Representation of the mPOFs proposed in: Refs. (a) [97,98], (b) [99], (c) [19], (d) [104] for the reduction of the scattering along the microstructured cladding during the POFBG inscription.

Regarding the work reported in Ref. [104], the authors proposed a new fiber design (see Figure 2.18d), capable to solve the drawbacks associated with the dopant diffusion that commonly occurs in core doped POFs, as well as to increase the POFBG writing efficiency. The proposed fiber was based on a double-clad layer, in which the inner cladding region with dimensions of ~19 to 24 μm, (for the 580 and 760 nm operational wavelengths, respectively), was made of TSB-doped PMMA material and an outer cladding region, made of undoped PMMA, with dimension of 125 μm. The doped region had an enhanced ultraviolet absorption [59] allowing to easily imprint POFBGs as reported in Ref. [42] and described in Subsection 2.5.4.1. The SM guidance was made by disposing one ring of air holes at the inner region of the TSB region, allowing to propagate the light at the central region (the dimensions of the core were ~2.6 to 3.4 μm, for the 580 and 760 nm operation wavelengths, respectively). By doing that, the influence of dopant diffusion during polymerization was minimized. Furthermore, the microstructure was composed of 3-air holes, separated by ~1.1 and ~1.6 μm (for the 580 and 760 nm operational wavelengths, respectively), allowing easy transverse coupling of the UV radiation to the central core region [104]. This interesting fiber configuration could potentially solve the two main drawbacks associated with the FBG recording in mPOFs which are the long exposure time and the grating reflectivity. However, no experimental results were reported to prove such theoretical predictions.

2.6.2 FBGs in Doped mPOFs

When fabricating SM SI-POFs one needs to balance the core radius and refractive index contrast between core and cladding. Considering PMMA as the fiber host material, the use of BzMA and ethyl methacrylate (EMA) monomers is commonly accepted in specific proportions for the refractive index modification of the core region. Nevertheless, dopant materials can be added to the polymer mixture, allowing to have fibers with photosensitive core materials, which benefits the POFBG inscription. However, the use of extra dopants in the core increases its refractive index, which needs to be compensated by the addition of other materials [105]. The use of mPOFs can, however, solve this drawback due to the absence of the manipulation of dopants and co-polymers to control the core refractive index. However, due to the low photosensitive of PMMA, it would still be interesting to dope the fibers with some photosensitive material for efficient photoinscription.

Reports on the inscription of FBGs in doped mPOFs can be found as early as 2006 [106,107]. The fibers were doped with TSB and FBGs were obtained through the phase mask technique and using a 325 nm CW HeCd laser. The gratings were written at the 1,550 nm window and had a peak-to-noise level from 8 to 15 dB. According to the authors, the benefit of the newly doped mPOF was the reduction of the inscription time to only 8 minutes when compared to the 15–20 minutes required to inscribe a POFBG in a similar fiber [106]. Despite the interesting results, no further information was given either about the fabrication of the fiber or on the structure type. Later, in 2013 and 2017, Sáez-Rodríguez et al. [77] and Hu et al. [78], respectively, have shown the possibility to dope the core of mPOFs with BDK, allowing to describe high reflective POFBGs (99.5% and 83.0%, respectively) with short periods of irradiation (13 minutes and 40 seconds, respectively). While both works presented in Refs. [77] and [78] share the same diffusion methodology in which a solution of methanol and BDK in the proportion of 3:1 is used to increase the mobility and deposition of BDK in PMMA, they are quite different regarding the mPOF production and inscription wavelength. Regarding the work developed by Sáez-Rodríguez et al. [77], the fiber was produced by doping a solid rod of PMMA in a solution of methanol and BDK for nine days. Next, the doped rod was introduced in the center hole of a preform containing a 3-ring hexagonal cladding as shown in Figure 2.19. The preform was then heat-drawn to a 130 µm fiber, which had mean holes diameter and pitch of 1.7 and 3.7 µm, respectively, giving a value of $d/\Lambda_d = 0.47$. The grating was then imprinted following the phase mask technique through a 30mW CW HeCd laser operating at 325 nm and using a phase mask with a period of 557.5 nm. The resultant grating was seen in transmission and it took about ~13 minutes to be written. It had 3.8 mm in length, a Bragg wavelength of 827 nm and showing a transmission depth of 23 dB [77].

Regarding the work developed later by Hu et al. [78], the fiber was produced using the selected center hole doping technique with a preform similar to the one used by Sáez-Rodríguez et al., composed of a central hole and surrounded by 3 air-hole layers. The holes surrounding the inner hole were blocked with UV resin and the preform was then immersed in a methanol-BDK solution. The solution was pulled to the inner hole by capillary action and then sealed with a film to prevent evaporation. The preform was then placed in a thermal chamber for 30 minutes at 52°C, allowing the solution to diffuse along the central air hole. The schematic of the process may be seen in Figure 2.20. The use of an oven at ~52°C allowed a diffusion process in about 30 minutes, which was faster than the one reported by Sáez-Rodríguez et al., who reported a nine-day period for doping the solid central hole rod at ambient temperature. After the doping process, the preform was heat-drawn to a fiber, where

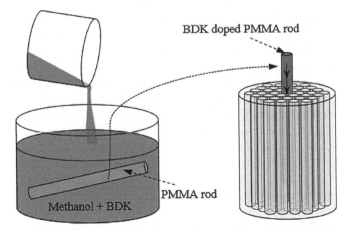

FIGURE 2.19 Schematic of the process used in Ref. [77] to dope the inner hole region of an mPOF preform by immersing a PMMA rod in a 3:1 methanol-BDK solution and filling it in the central hole of the undoped PMMA mPOF preform.

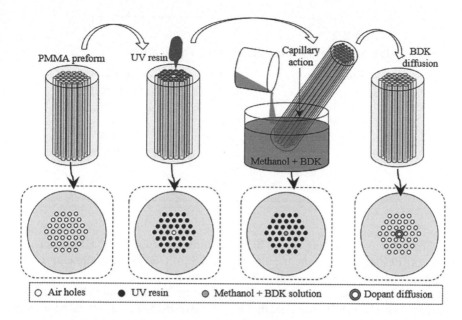

FIGURE 2.20 Scheme used in Ref. [78] for the central hole doping process in a microstructured preform. In the first stage the air holes are blocked with UV resin (except for the central air hole), then the preform is immersed in a 3:1 methanol + BDK solution to fill the central hole by the capillary action followed by dopant diffusion.

the central hole is collapsed due to the low melting temperature of BDK compared to PMMA. The final fiber had a diameter of 150 μm and was composed of a solid and smooth central core, being the dimensions of the hole diameter and pitch equal to 1.5 and 3.8 μm, respectively, giving a hole to pitch ratio of 0.4. The only drawback of the proposed scheme was the higher loss of the drawn fiber at the 800 nm region, which was 100 times higher than the one obtained for a similar undoped PMMA fiber. After fiber fabrication, the gratings were written through the phase mask technique using a femtosecond laser operating at the 2nd harmonic (400 nm), having 11 mW power and using a phase mask with a period of 1,060 nm. This approach resulted in FBGs inscribed in 40 seconds, with 10 mm length, located at ~1,565 nm and with transmission depth of ~7.7 dB [78].

2.6.3 FBGs IN mPOFs USING 248 nm UV LASER

As early as 2001, when the first FBGs in SI-POFs were being reported, the increase of the intensity of the UV laser beam at the core POF through the use of a pair of lens for the reduction of the laser beam spot from 6 mm to about 2 mm in the transversal direction of the fiber length, was described to reduce the inscription time from 85 minutes to ~15 minutes [108]. A decade later, Bundalo et al. explored also the property of increasing the power reaching the core POF to report a fast inscription of POFBGs in mPOFs in just 7 minutes [15]. Other authors also tried to reduce the amount of time needed to inscribe an FBG, such as through the orientation of the microstructured

TABLE 2.2

UV Sources Used for the POFBG Fabrication

Wavelength (nm)	Laser	Reference
248	KrF	[12,27]
	OPO	[39]
280	OPO	[39]
266	Nd:YAG	[80]
308	XeCl	[55,109]
325	OPO	[3,39]
	HeCd	[6,54,68]

cladding holes to the writing beam [103], use of femtosecond lasers [19], absence of air holes [19,104] and also the use of core doped mPOFs [78,95], allowing in this case, the capability to achieve high reflective gratings [78,95].

Overall, most of the works regarding Bragg grating inscription in POFs used HeCd lasers operating at 325 nm wavelength. Deeper wavelength sources would be preferable due to the enhanced photosensitivity. However, the attenuation of polymers at those wavelengths was soon realized to be the major problem for the Bragg grating inscription in POFs. Peng et al. showed the first unsuccessful results on the inscription of Bragg gratings on a core PMMA preform using the 248 nm wavelength [3]. The results revealed a surface relief grating, which was attributed to the higher attenuation of the polymer at the 248 nm wavelength region. Yet, they showed that a 325 nm wavelength source could be used instead, to create a bulk core refractive index modification [3]. Since then, this wavelength source was the preferred choice for POFBG fabrication.

However, works performed by other authors showed that deep wavelengths could also be used for the inscription of POFBGs. Examples of those wavelength sources may be seen in Table 2.2.

The efficiency of the POFBG inscription using these UV sources has been mainly attributed to different mechanisms such as photodegradation, photopolymerization, photocrosslinking, photolocking, photoablation, and so on (see Section 2.2). Furthermore, the refractive index modification in some of those works can achieve index modulations of the order of 10^{-4}.

Fiber Bragg gratings in silica fibers are well established and because of that, fiber Bragg grating recording systems are already developed and spread worldwide. Most of them rely on the use of KrF UV lasers which operate at 248 nm wavelength, known to provide higher photosensitivity for both silica and polymer materials. Therefore, the possibility to use POFs in these widely available inscription systems is of great interest.

2.6.3.1   Bragg Grating Fabrication in a Few-Mode mPOF

Unsatisfied with the tens of minutes required to inscribe a POFBG using 325 nm HeCd sources and taking into account the literature reports on the capability to change the refractive index of undoped PMMA films (see Section 2.2), through the use of a 248 nm KrF laser, Oliveira et al. [50] decided to study the POFBG inscription under low frequencies (1 Hz), low fluence (33 mJ/cm²), and low exposure time (<30 seconds), avoiding the regime of polymer ablation. Their results proved that under those conditions, it was possible to imprint high-quality FBGs in undoped mPOFs in a record inscription time of 20 seconds [50], paving the way for a more intense use of the POFBG technology, either in part due to the low inscription time, as well as due to the possibility to use the already installed silica FBG recording systems. Due to the significance of this result, it will be presented in the next paragraphs.

The tests performed in Ref. [50] were done by implementing a KrF Bragg Star™ Industrial-LN excimer laser operating at 248 nm. The laser beam had a spot size of 6 mm in width and 1.5 mm in height, with a pulse duration of 15 ns. The fiber under experiment was a commercial one, purchased from the former Kiriama Pty Ltd. and it provided FM behavior at the visible region. It was made of undoped PMMA and it was composed of six layers of air holes surrounding a solid inner core. The cladding diameter of the fiber was around 250 μm, the holes diameter measured 3.2 μm and the hole-to-hole separation was about 6.2 μm. A microscope image of the fiber may be seen in Figure 2.21a, while the near field image of some of the modes allowed to propagate into the fiber, may be seen in Figure 2.21b–e.

In a preliminary test, the authors decided to investigate the laser repetition rate and fluence, ensuring that the surface of the mPOF was not damaged/ablated. Thus, a special inscription setup was prepared, which is represented by the schematic shown in Figure 2.22. The setup was composed of two mechanical stages to secure the fiber and an optical apparatus composed by mirrors, lens and an adjustable slit that allowed to guide, focus and adjust the length of the laser beam onto the mPOF, respectively. For the first trial, the laser repetition rate (R) was set to 50 Hz, and the pulse energy (E) set to 3 mJ, giving a fluence of 33 mJ/cm² per pulse, which was well below the threshold reported for PMMA ablation [27]. The irradiation was performed in few seconds and it was concluded that R was too high because the surface appeared damaged by naked eye. The same conclusion was obtained for the same fluence but for $R = 5$ Hz during 60 seconds. It was observed that there was still damage to the fiber surface as shown in microscope image shown in Figure 2.23a. A conclusion taken from the results was

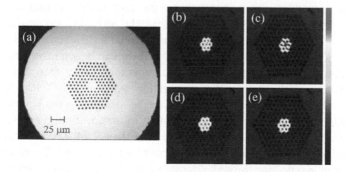

FIGURE 2.21 (a) Microscope image of the FM-POF used in Ref. [50]; (b)–(e) near field image of some of the modes allowed to propagate in the fiber.

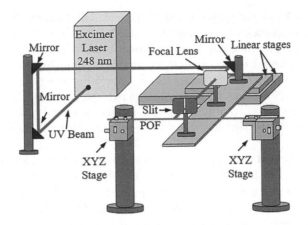

FIGURE 2.22 Setup used in Ref. [50] for the POF irradiation with a 248 nm KrF UV radiation. After the beam exit, the UV radiation is guided by the reflection of three sequential mirrors. The UV beam goes through a lens and slit which then shapes the beam onto the POF.

FIGURE 2.23 Microscope images of the surface of the FM-mPOF exposed to 248 nm UV radiation, with a repetition rate of (a) 5 Hz and (b) 1 Hz, for the same fluence of 33 mJ/cm², and exposure time of 60 seconds. The image shown in (a) clearly shows surface ablation. (Reprinted from R. Oliveira, et al., "Bragg gratings in a few mode microstructured polymer optical fiber in less than 30 seconds," *Opt. Express*, vol. 23, no. 8, pp. 10181–10187, 2015, with permission from the Optical Society of America.)

that the damages on the surface fiber were mainly due to the rise of temperature on the irradiated region, which was a consequence of the laser frequency. Therefore, the laser parameters were set to its minimum, with $R = 1$ Hz, and $E = 3$ mJ, during the same period of time. By doing that, the cumulative heating on the fiber was minimized as well as the occurrence of ablative processes [110]. As can be seen from Figure 2.23b, no ablation on the fiber surface was observed, providing clear evidences that R and E were suited for the trials related with Bragg gratings inscription in POFs using 248 nm.

Based on the preliminary results, Oliveira et al. decided to implement the phase mask technique, by placing a phase mask close to the focal point of the UV-beam and in the close vicinity of the fiber, as shown in Figure 2.24. The phase mask had a uniform period of 1,023 nm and it was designed for the 248 nm. The grating growth was monitored in reflection through a commercial fiber interrogator and the coupling of the POF to silica SM fiber, made through the help of a side camera and a laser beam profiler, as explained in Subsection 1.4.2.

FIGURE 2.24 Picture of a 248 nm phase mask inscription setup and aligning system used in Ref. [50]. Legend: "*A*" → KrF laser (not shown); "*B*" → beam profiler; "*C*" → 40 X objective; "*D*" → motorised linear stage (M-ILS150CC); "*E*" → mirror; "*F*" → Lens; "*G*" → Slit; "*H*" → Phase mask; "*I*" → POF; "*J*" → Butt coupling; "*K*" → telecentric lens coupled to camera; "*L*" → silica pigtail fiber; "*M*" → 3D mechanical axis; "*N*" → linear stage (ABL20100).

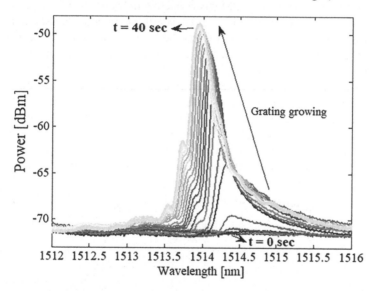

FIGURE 2.25 Bragg grating reflection spectra of a FM-250 mPOF acquired with a frequency of 1 Hz, for the inscription with a 248 nm KrF laser, with inscription parameters of: $L = 4.5$ mm, $\Lambda_{pm} = 1,023$ nm, $R = 1$ Hz, fluence of 33 mJ/cm², and number of pulses to reach saturation of $N = 20$ pulses. (Adapted from R. Oliveira, et al., "Bragg gratings in a few mode microstructured polymer optical fiber in less than 30 seconds," *Opt. Express*, vol. 23, no. 8, pp. 10181–10187, 2015, with permission from the Optical Society of America.)

The first results showed that the inscription of FBGs in the FM-mPOF with the 248 nm UV laser could be made in few seconds, as shown Figure 2.25, revealing an unprecedented result considering the tens of minutes reported for the conventional 325 nm UV radiation [3,6].

The results revealed that the Bragg peak appeared at the first irradiation pulse, growing at a high rate until 20 pulses (20 seconds). Despite the fiber being FM, the spectra appeared with a single Bragg peak at ~1,514 nm. The result was however related to the axial alignment made between the silica SM fiber to the FM-mPOF, which preferentially couple to the fundamental mode of the fiber. The peak power and the Bragg wavelength evolution were also monitored and can be seen in Figure 2.26a and b, respectively.

From Figure 2.26a, one can observe that the Bragg peak reached saturation in around 20 seconds, meaning that only 20 pulses were needed to create an index modulation in the core POF, giving total cumulative energy of 60 mJ. Furthermore, the results collected for the Bragg wavelength shown in Figure 2.26b, revealed a blue-shift during the writing process, maintaining its value after the irradiation was off. The phenomenon was similar to the ones observed for the inscription of POFBGs with 325 nm laser, where the blue-shift is due to the combination of refractive index change and rise of the temperature due to the UV absorption, where the latter revealed to be a reversible process, due to the red wavelength shift after the laser is switched off. Judging the results, one may assume that the low repetition rate gave enough time to the fiber to cool down between consecutive irradiation pulses. According to Equation 2.9, and considering: a PMMA refractive index of 1.4794, calculated from the Sellmeier equation (Equation 1.11) and using the coefficients presented in [111]; a Bragg wavelength of 1514.43 nm for the first moment when the Bragg peak appears; and 1513.95 nm for the Bragg wavelength at the end of the irradiation process, it can be calculated a core mean refractive index equal to -4.7×10^{-4}. One interesting result is the negative value of the refractive index change. In literature, it has been reported both positive and negative values in PMMA slabs under 248 nm UV radiation [23,30]. The explanation has been essentially described as the result of photochemical reactions, rather than photo-thermal [26]. Furthermore, literature results also demonstrate that PMMA ablation does not start with either the first pulses, even considering a wide fluence range, or with radiation above 270 nm [26]. The ablation only occurs after a

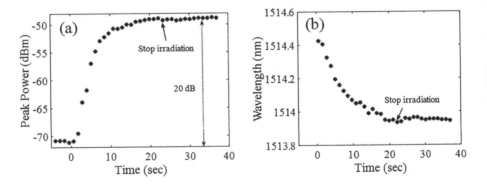

FIGURE 2.26 Evolution of the Bragg peak power (a) and Bragg peak wavelength (b) of an FBG written in an undoped PMMA mPOF using a 248 nm KrF laser, with inscription parameters of $L = 4.5$ mm, $\Lambda_{pm} = 1,023$ nm, $R = 1$ Hz, fluence of 33 mJ/cm^2. (Reprinted from R. Oliveira, et al., "Bragg gratings in a few mode microstructured polymer optical fiber in less than 30 seconds," *Opt. Express*, vol. 23, no. 8, pp. 10181–10187, 2015, with permission from the Optical Society of America.)

required certain number of incubation pulses. Additionally, Burns et al. have shown that the threshold for ablation decreases with the repetition rate, being that a result of the heating on the sample due to the sequential laser pulses [110]. On the other hand, Küper and Stuke revealed that under the incubation phenomenon the UV absorption of PMMA samples increases with the number of pulses, obtaining transmissions of less than 6% for a 40 μm sample irradiated under 248 nm with 1,000 pulses and with a fluence of 35 mJ/cm², indicating significant changes in the chemical constitution of the material [26]. Detailed information of those species was obtained from infrared spectra, which revealed the presence of unsaturated species. Further studies provided by Wochnowski et al. showed that using the same wavelength light and at relatively low fluence (15 mJ/cm²), lead to a complete side chain separation from the main chain (see Figure 2.1, (ester main-chain scission)) generating free methyl formate radical that was detectable through X-ray photoelectron spectroscopy (XPS) and quadru-pole mass spectroscopy (QMS) measurements. They have also shown that continuous irradiation at fluence of 30 mJ/cm² or more lead to a degradation of the free ester radical either into methane and carbon dioxide detected by XPS measurements or into carbon monoxide and methanol as indicated by Fourier transform infrared spec-troscopy (FTIR) and QMS measurements [30]. The complete separation of the side chain of PMMA causes the formation of a radical electron at the α-C-atom, which leads to a destabilization of the polymer main chain leading to its scission at high irradiation doses and the formation of a C=C double bond as identified in the FTIR measurements in Ref. [30]. In view of the previously reported literature results, one can conclude that the drop of refractive index is explained essentially as a result of the degradation of the main polymer structure as already concluded in Refs. [24,25]. Thus, shorter molecule chains are generated, leading to a higher degree of freedom, leading to a decrease in density and inherently in the refractive index. The conclusion was thus similar to the ones reported for POFBGs inscribed with 325 nm UV lasers.

2.6.3.2 Fiber Bragg Grating Fabrication in Different mPOFs

The success on the work presented by Oliveira et al. related to the inscription of an FBG in a FM-mPOF in few seconds through the use of a 248 nm UV laser [50], led to subsequent works, related to the inscription of FBGs in different mPOFs, either SM, FM, MM, highly-birefringent (HiBi) or doped ones [86,102,112,113]. An exam-ple of those works may be seen in Figure 2.27 for four different mPOFs shown in Figure 2.27a, written under the same conditions as reported in Ref. [50] and using a phase mask with a period of 1,033 nm. The results concerning the peak power reflec-tivity during inscription time may be seen in Figure 2.27b and the Bragg reflection spectra after inscription process in Figure 2.27c–f.

The mPOFs shown in Figure 2.27a were acquired from the former Kiriama Pty Ltd and are SM for the "SM-125," FM for the "FM w/Rh6G," and MM for the "MM-G3" as well as the "MM-150." Furthermore, they were made of undoped PMMA, except for the "FM w/ Rh6G," which was doped with Rhodamine 6G (Rh6G).

From Figure 2.27b, it can be seen that regardless of the fiber dimensions, hole lay-ers and dopants, all the gratings reached their saturation level in less than 30 seconds, similar to the one found in Ref. [50]. Regarding the reflection spectra shown in

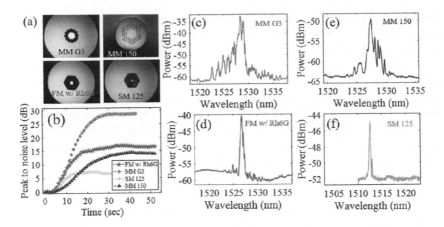

FIGURE 2.27   (a) Microscope images of four different types of mPOFs (MM G3, MM 150, FM doped with Rh6G and SM 125). (b) Peak-to-noise level evolution during inscription time. (c)–(f) Spectra acquired after the POFBG inscription for different mPOFs.

Figure 2.27c–f, they show a multi-peak behavior in (c) and (e) due to the MM nature of the fibers; few reflection peaks in (d) due to the less number of modes propagating into the POF; and a single peak for the fiber presenting SM behavior, shown in (f). Nevertheless, it is worth to mention that the spectra for the FM and MM fibers also depends on the coupling condition, in which preferential modes could be excited regarding the transversal orientation of the silica fiber to the mPOFs. One interesting finding from the results shown in Figure 2.27b is the influence of the UV scattering effect on the number of air hole layers which has an impact on the reflectivity of each grating. In particular, MM-G3 presents the best peak-to-noise level (~30 dB), for just a single air hole layer, when compared with the lower peak-to-noise level observed for the SM-125 (only 7 dB) for a microstructured cladding composed of six air hole layers. Regarding the later, it would be a good opportunity to explore the irradiation under the ΓK orientation, as explored by Marshal et al. [101] in PCFs and later by Bundalo et al. in mPOFs [103]. Such approach would allow the UV radiation to reach the core more efficiently and thus, getting strong refractive index modulations due to the less scattering on the microstructured region and inherently higher POFBG reflectivity. Nevertheless, the peak-to-noise level of the gratings shown in Figure 2.25 and Figure 2.27 are well accepted for most of the peak detection algorithms found in most FBG sensor demodulation schemes.

In 2013, Saéz Rodríguez at al. showed that using PMMA-mPOF doped with BDK was a viable alternative to reduce the inscription time to about 13 minutes using the UV radiation from a 325 nm CW HeCd laser [77]. Later Hu et al. showed that a time reduction to about 40 seconds was possible by using the 2nd harmonic of a fs laser, at 400 nm [78]. However, these wavelengths sources were still far from the peak wavelength absorption located at around 230–270 nm [56]. Therefore, the use of a more appropriate laser source could provide a reduction on the time associated with the inscription process. In this way, and taking into account the results reported by Oliveira et al. [50] for the POFBG inscription with 248 nm KrF UV pulsed laser in

TABLE 2.3

Evolution of the Exposure Time Needed to Inscribe FBGs in mPOFs, Using Different Inscription Procedures

Exposure	Laser Source	λ_{Bragg} (nm)	mPOF	Material	Dopant	Year	Ref.
60 minutes	325 nm CW HeCd	1,570	SM	PMMA	-	2005	[6]
43 minutes	325 nm CW HeCd	658	SM	PMMA	-	2013	[16]
13 minutes	325 CW HeCd	827	SM	PMMA	BDK	2013	[77]
8 minutes	325nm CW HeCd	1,572	SM	PMMA	TSB	2006	[106]
7 minutes	325 nm CW HeCd	633	SM	PMMA	-	2014	[15]
6 minutes	325 nm CW HeCd	892	SM	PC	-	2016	[8]
5 minutes	325 nm CW HeCd	866	SM	ZEONEX® 480R	-	2017	[10]
40 seconds	400 nm fs laser	1,565	SM	PMMA	BDK	2017	[78]
30 seconds	248 nm KrF pulsed	1,514	FM	PMMA	-	2015	[50]
2.5 seconds	800 nm fs laser (PbP)	1,519	SM	PMMA	-	2012	[19]
15 nanoseconds	248 nm KrF pulsed	841	SM	PMMA	BDK	2017	[79]
8 nanoseconds	266 nm Nd:YAG pulsed	841	SM	PMMA	BDK	2018	[80]

undoped PMMA-POFs, Pospori and co-workers [79] used the same writing system with the doped mPOF fabricated by Saéz Rodríguez at al. found in Ref. [77]. The results showed the capability to write a POFBG centered at 850 nm region with a single KrF UV laser pulse (pulse duration of 15 ns) [79]. Later, they used a Nd:YAG laser, operating at 266 nm, with pulse duration of 8 nanoseconds and pulse energy of 72 µJ, together with the BDK doped mPOF produced by Hu et al. found in [78], showing also the possibility to write an FBG with one single pulse [80].

Back in time, we have seen a huge improvement in the exposure time needed to inscribe POFBG in mPOFs, which decreased from hours [6] to just a few seconds. Table 2.3 gives an insight into this evolution for some of those works. Different methodologies, laser power, materials, dopants, laser sources, etc., have been used in different works and nowadays the duration of the inscription of POFBG is no longer an issue. Therefore, we may say that the POFBG fabrication technology has matured and the possibility to inscribe FBG during the POF drawing process is now a reality.

2.6.4 FABRICATION OF FBGS IN MPOFS MADE OF SPECIAL POF MATERIALS

Doped mPOFs have shown their potential to produce high-quality FBG with low exposure periods. However, the fabrication of those fibers requires complex procedures [77,78] and because of that it is more expensive than standard mPOFs. Furthermore, they have shown to provide higher losses [78] and are also unsuitable for in-vivo applications.

Nowadays, the availability of different polymer materials such as TOPAS®, PC, ZEONEX®, etc., together with the development of mPOFs based on them, opened the possibility to explore a variety of physical and chemical properties. The inscription of FBGs on those fibers not only revealed to be possible, but also provided the

TABLE 2.4

FBGs in Different mPOF Materials and Their Inherent Properties

Material	d/Λ_d	Property	POFBG Region (nm)	Reference
TOPAS® 8007	0.45	RH insensitive	~1,550	[99]
TOPAS® 5013	<0.43	High T_g; RH insensitive	~850	[114]
PC	0.40	High T_g	~850	[8]
ZEONEX® 480R	0.40	High T_g RH insensitive	~850	[10]
ZEONEX® 480R + PMMA	0.42	Simultaneous detection of RH & temperature	~850	[115]

fuel for a range of sensing applications not available before, such as the ones with humidity insensitiveness [10,99,114], high thermal resistance (125°C) [8,10,114], easy fabrication [10], and capability to simultaneously measure RH and temperature [115]. Table 2.4 contains an overview of the properties of those POFBGs.

The first report on the inscription of FBGs in mPOFs made of materials other than PMMA was published in 2008 for a TOPAS grade 8007 [116]. The motivation to use TOPAS® is driven by its lower moisture absorption, which is defined to be a hundred times lower than that of PMMA, allowing to create humidity insensitive sensors. Contrary to PMMA, this material is chemically inert and the direct binding of biomolecules onto the surface is difficult. Nevertheless, reports on the use of antraquinon linker molecules to attach the TOPAS® surface when activated by UV radiation, have been proposed for posterior acceptance of sensor layers [116]. The fiber produced in Ref. [116] was composed of a 400 μm cladding diameter and an 80 μm core surrounded by two layers of air holes with a diameter of 55 μm and disposed in a hexagonal configuration. The Bragg grating was written with the phase mask technique using a 30 mW CW HeCd laser operating at 325 nm and a phase mask period of 1057.2 nm. The grating was strangely observed in transmission but not in reflection, comprising a resonance Bragg wavelength centered at 1,600 nm and having a surprisingly positive response with temperature [116]. However, according to a later study provided by the same group, it was claimed that the results were not reproducible [7]. Their findings were made for the same fiber material, but considering a 270 μm SM mPOF composed of two air layers disposed hexagonally and having a hole diameter and hole-to-hole distance of 3.8 and 8.5 μm, respectively. The grating was reported using the same inscription procedure and with a phase mask period of 1034.2 nm, allowing to obtain a Bragg peak after 45 minutes and centered at 1,570 nm [7]. The fiber was then characterized to temperature, showing a reproducible negative tendency with a sensitivity of nearly 37 pm/°C. Later reports on similar mPOFs revealed the possibility to write Bragg gratings at the 850 nm region, where the polymer losses are lower [99]. Characterization of the fiber to humidity, revealed an almost insensitive nature, showing a residual sensitivity of 0.3 ± 0.1 pm/%RH.

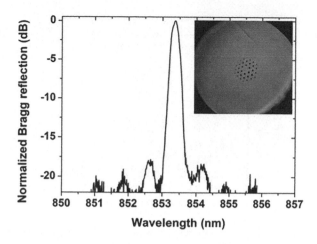

FIGURE 2.28 FBG recorded in TOPAS® 5013 mPOF. The inset shows the cleaved end face mPOF. (Reprinted from C. Markos, et al., "High-Tg TOPAS microstructured polymer optical fiber for fiber Bragg grating strain sensing at 110 degrees," *Opt. Express*, vol. 21, no. 4, pp. 4758–4765, 2013, with permission from the Optical Society of America.)

The inscription of FBGs in mPOFs based on a different grade of TOPAS®, at the same spectral region and using the same phase mask, was later reported by Markos et al. for a TOPAS® grade 5013 [114], the result may be seen in Figure 2.28.

The grating recorded in Ref. [114] was developed in order to show humidity insensitivity, as the one recorded in TOPAS® 8007 [7], while being still capable to sense strain at high temperature (up to 110°C), due to the higher glass transition temperature (T_g) presented for the grade 5013, compared to 8007 (i.e., 80°C vs. 135°C, respectively). However, Woyessa et al. reported that TOPAS® 5013 fibers are difficult to draw, due to the low molecular weight [10]. According to the authors, polymers with similar physical characteristics such as ZEONEX® 480R are more suited due to their high molecular weight (i.e., 480,000 g/mol), which is six times higher than the one presented for TOPAS® 5013. In view of that, a wide range of drawing stress provides an easier tuning of the final mechanical properties of the fiber. Furthermore, the higher degree of freedom in fiber design allows to get structures such as air holes in the cladding region with minor distortions [10]. Based on the similar physical characteristics of ZEONEX® 480R, compared to TOPAS® 5013 (either on its T_g and low water absorption), and taking into account the easiness on the drawing process of the previous, Woyessa et al. reported the fabrication, Bragg grating inscription and characterization of a ZEONEX® 480R mPOF at the 850 nm region. The result may be seen in Figure 2.29.

The mPOF produced by Woyessa and co-workers and shown on the inset of Figure 2.29, had a cladding diameter of 150 μm, hole diameter of 2.2 μm, and a hole to hole separation of 5.5 μm, allowing to have $d/\Lambda_d = 0.4$, to permit endlessly SM operation. The grating was written at the 850 nm region, where the fiber presented a minimum attenuation of ~3 dB/m. For that, the phase mask technique was used through a CW HeCd laser, using a phase mask period of 572.4 nm and attenuating

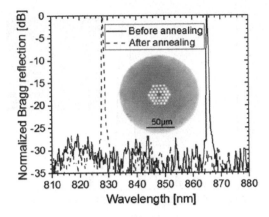

FIGURE 2.29 FBG recorded in ZEONEX® 480R mPOF, before and after the annealing process. The inset shows the end face mPOF. (Reprinted from G. Woyessa, et al., "Zeonex microstructured polymer optical fiber: Fabrication friendly fibers for high temperature and humidity insensitive Bragg grating sensing," *Opt. Mater. Express*, vol. 7, no. 1, pp. 286–295, 2017, with permission from the Optical Society of America.)

the 2 mm UV beam to 5.5 mW, allowing to get a grating in 5 minutes as shown in Figure 2.29 (either before or after the annealing process). The characterization results revealed: a similar strain sensitivity to the ones found for fibers made of different materials; operation at high temperatures (i.e., 100°C); as well as humidity resistance (sensitivity of 0.8 ± 0.4 pm/%RH) [10].

The good compatibility of ZEONEX® and PMMA has also been explored by Woyessa et al. for the fabrication of a POFBG sensor capable to simultaneously measure humidity and temperature [115]. The result was made possible through the development of a ZEONEX® 480R mPOF, having an overlay thickness of ~25 μm. The gratings were written through the same phase mask in the fiber with and without the PMMA overlay. By doing that, different responses to humidity and temperature were achieved, and a dual parameter sensor has been developed [115].

FBGs recorded in mPOFs made of TOPAS® 5013 [114] and ZEONEX® 480R [10], presented higher operation temperature when compared to PMMA or TOPAS® 8007, due to their higher T_g (134°C and 138°C > 109°C and 78°C, respectively). Nevertheless, PC, another thermoplastic material, presents good mechanical properties (yields and breaks at elevated values) and it has one of the highest T_g among the transparent plastics (i.e., ~145°C). Because of these properties, in 2016, Fasano et al. reported the first PC mPOF [8]. The drawn fiber, shown in the inset of Figure 2.30, had an outer diameter of 150 μm and was composed of three rings of air holes disposed in an hexagonal arrangement, with diameter of 1.8 μm and separated by a distance of 4.4 μm, allowing to get endlessly single-mode operation ($d/\Lambda_d = 0.4$). The fabricated mPOF had lower losses at the 850 nm region, reaching values of ~9 dB/m. In this way, in order to show the opportunities of the fiber for sensing applications, the authors reported the inscription of a POFBG at this spectral region. The inscription methodology was the one found in Refs. [114,115], being the phase mask pitch equal to 572.4 nm and the laser power set to 4 mW. The grating was inscribed in 6 minutes,

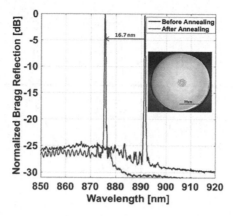

FIGURE 2.30 FBG inscribed in PC mPOF, before and after the annealing process. The inset shows the end face of the cleaved mPOF. (Reprinted from A. Fasano, et al., "Fabrication and characterization of polycarbonate microstructured polymer optical fibers for high-temperature-resistant fiber Bragg grating strain sensors," *Opt. Mater. Express*, vol. 6, no. 2, pp. 649–659, 2016, with permission from the Optical Society of America.)

having a peak-to-noise level of ~25 dB and located at around 890 nm, which was then blue-shifted to ~875 nm, after a post-annealing process, as it is shown in Figure 2.30.

The annealing of the POFBG was made in a two-step procedure, either at 120°C during 24 hours and then at 130°C for 12 hours. By doing that, the authors were then able to characterize the shift of the Bragg wavelength peak with increasing and decreasing temperature up to a maximum of 125°C, which is the highest value reported with a POFBG. Such result opens new perspectives on the use of POFBGs, such as in sterilization in biomedical applications, which normally require steam autoclaving at a temperature of ~120°C [8]. However, it is also worth to mention that PC presents high moisture absorption capabilities and thus, POFBG sensors based on this type of fiber material should take into account the cross-sensitivity to humidity.

TOPAS® mPOFs [7,99,114], as well as the ones based on ZEONEX® 480R [10] and PC [8], revealed to be intrinsically photosensitive under UV radiation, namely for the CW HeCd laser operating at 325 nm [7,8,10,99,114]. Furthermore, they showed inscription times ranging between a few minutes in the case of PC and ZEONEX® 480R, to hours in the case of TOPAS® 8007. Based on these characteristics and taking into account that the inscription time could be shortened to less than 30 seconds using a pulsed KrF UV laser operating at 248 nm, as demonstrated by Oliveira et al. for different undoped PMMA mPOFs [50,112]. The same system was used to obtain gratings in few seconds on new mPOF materials [117].

2.7 FIBER BRAGG GRATING FABRICATION AT THE LOW LOSS REGION OF POLYMERS

Historically, the qualities of polymer optical fibers have been overwhelmed by the popularity of the silica optical fibers. This has been mainly due to the POFs higher transmission loss. In fact, the loss mechanisms affect silica fibers as well as POFs. However, this

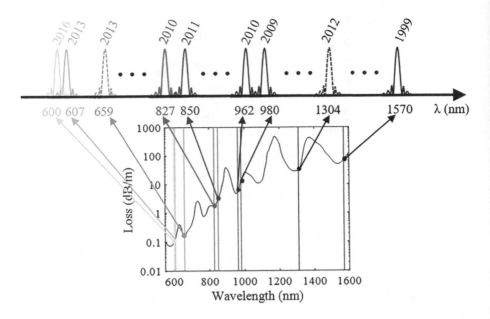

FIGURE 2.31 Evolution of the resonance Bragg wavelength in POFs along the years, achieved by different authors, namely for the wavelengths of 1,570 nm [3]; 1,304 nm [119]; 980 nm [55]; 962 nm [45]; 850 nm [54]; 827 nm [100]; 659 nm [16]; 607 nm [14]; 600 nm [118]. The graph corresponds to the PMMA attenuation, where it can be seen a loss improvement for the short wavelength region.

is much more pronounced for POFs due essentially to the overtones of the molecular vibration absorption of the basic elements of the polymer material (see Section 1.2.4).

The visualization of the FBG signal is inherently affected by the attenuation of the material. For POFs, the attenuation of the majority of polymers at the infrared region becomes so high that it limits the length of POFBG devices to just a few centimeters. Therefore, the applicability of POFBG technology could be very limited. Thus, different groups ([14,16,45,54,55,100,118,119]) took their efforts to shift the resonance Bragg wavelength to the near-infrared and visible regions. A schematic representation of this evolution along the years, considering the corresponding PMMA material loss, may be seen in Figure 2.31.

2.7.1 POFBGs AT THE NEAR-INFRARED REGION

The first report of a POFBG was made for a PMMA fiber at the 1,570 nm region in 1999 [3]. After that, different authors have written Bragg gratings at this spectral region in POFs of different materials and structures [6,42,56,98]. Despite the high losses at this spectral region, the authors were mainly motivated by the availability of light sources and detectors which are well established for the silica fiber telecom industry. However, in most practical applications, a longer POF length is more desirable. Hence, an FBG at shorter wavelengths, where polymers offer lower losses is preferred. Following that goal, in 2009, Terblanche et al. showed the first results of

POFBGs at shorter wavelengths, namely at ~980 nm [55], triggering the first steps for low loss POFBG sensor devices. The gratings were written in two PMMA step-index SM-POFs being the core of one of those fibers composed of PMMA-co-PS while poly(MMA-co-EMA-co-BzMA) for the other. The inscription followed the phase mask technique through a XeCl excimer laser operating at 308 nm with a repetition rate of 8 Hz and fluence of 70–85 mJ/cm². The phase mask period had 658 nm which allowed them to get a resonance Bragg wavelength centered at 980 nm and reflectivity between 8% and 24%. However, for unknown reasons, the authors report that those gratings completely disappeared one to two days after the inscription [55]. One year later two other authors, namely, Johnson et al. [100] and Zhang et al. [45], showed stable POFBGs at 827 and 962 nm, respectively. The motivation behind the work reported by Johnson and co-workers [100], which reports a shorter Bragg wavelength than the ones reported by Terblanche et al. [55] and Zhang et al. [45], was due to a better attenuation at the 830 nm region (fiber loss of ~2 dB/cm), which according to the authors was very convenient due to the availability of broad-band semiconductor sources at that wavelength [100]. Furthermore, their inspiration on the possibility of achieving refractive index modulations with shorter periods was based on the work developed in the 1970s for recording resolutions better than 200 nm in a PMMA bulk sample [120]. Based on these motivations, Johnson and co-workers were able to imprint an FBG centered at 827 nm wavelength in a PMMA MM-mPOF using the phase mask technique, through a CW 325 nm HeCd laser with output power of 30 mW [100]. A case example of gratings written in the same conditions as described before, may be seen in Figure 2.32 for an FBG written in a PMMA MM-mPOF, using a phase mask with period of 572.12 nm.

The problem of FBGs in MM fibers regarding sensing applications relates to their multi-peak nature, which creates difficulties on the Bragg wavelength tracking, due to the power exchange between modes that satisfy the Bragg condition. Additionally,

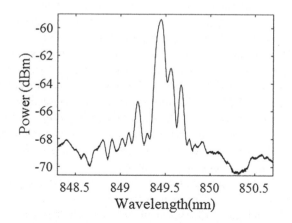

FIGURE 2.32 POFBG at the 850 nm region, obtained through the phase mask technique using a 30 mW power HeCd laser, through phase mask with a period of 572.12 nm, in a PMMA MM-mPOF (G3-mPOF fiber sold from the former Kiriama Pty Ltd.). It is possible to observe multiple peaks at around 849.5 nm due to the MM nature of the POF.

the bandwidth of those gratings is normally high, (i.e., a bandwidth of ~2.45 nm [100]), which results in poor resolution on the Bragg peak detection, and inherently on the parameter being measured. Because of that, in 2011, it was reported the fabrication of FBGs in the same spectral region but using a PMMA-based POFs with FM behavior, either SI (core composed of PMMA-co-PS) or also mPOF, through a phase mask with period of 572.4 nm, obtaining resonance Bragg wavelengths with narrow bandwidths of 0.17 and 0.29 nm, respectively [54].

2.7.2   POFBGs at Short Wavelengths

Despite the efforts in showing POFBGs at the 850 nm region, for MM-POFs in 2010 [100], and FM-POFs in 2011 [54], the minimum attenuation for PMMA material is located at around 600 nm (see Figure 2.31). In those days SM-POFs were already available by the former Kiriama Pty Ltd. and Paradigm Optics Inc., where fiber losses at the 600 nm region could achieve values as low as 1.0 and 0.2 dB/m, respectively, [14]. Therefore, in 2013, two different groups showed the first POFBGs inscribed at the 600 nm window [14,16]. Both of them used the phase mask technique through a 30 mW CW HeCd laser operating at 325 nm. However, the period of the phase masks used by them was different, namely: 405.4 nm [14] and 855 nm [16]. Regarding the work developed in Ref. [14], it was reported the fabrication of 1 cm gratings through the scanning technique in mPOFs made of PMMA, either MM or FM, allowing to obtain Bragg gratings centered at 605–607 nm (matching the theoretical value given by the Bragg condition, considering a refractive index at this spectral region of around 1.49). Inscription of Bragg gratings at this spectral region using phase masks with similar pitches were then followed by other authors [15,103].

Phase masks are designed to provide maximum contrast for the interference of the ±1 diffraction orders, through the suppression of the zeroth and higher diffraction orders. However, the lower efficiency of the zeroth diffraction order can, however, interfere with the ±1 diffraction orders, leading to the formation of a Talbot interference pattern behind the phase mask [43], giving rise to intensity variations with a periodicity of $\Lambda = \Lambda_{pm}$. These type of periodicities have been visually confirmed in either silica [121], or polymer fibers [43], being an example of such phenomenon shown in Figure 2.33 for a POFBG written with a phase mask period of 1,033 nm with ±1 principal orders, and with 2.4% efficiency on the zeroth-order.

As shown in Figure 2.33, the distance between two consecutive fringes is about 2,070 nm, indicating a period of ~1,035 nm, which is similar to the period of the phase mask (1,033 nm), being the error given by the measurement resolution. This phenomenon has already been reported in silica fibers to give Bragg gratings with resonances located at λ_{Bragg} and $2\lambda_{Bragg}$, directly related to the periods of $\Lambda/2$ and Λ. Moreover, it has also been demonstrated for both silica [121] and polymer fibers [16] that the presence of higher diffraction orders gives rise to Bragg reflection peaks located at $\lambda_{Bragg}/2$ and $2\lambda_{Bragg}/3$. Based on this knowledge, Statkiewicz-Barabach showed the fabrication of POFBGs at different spectral regions, using a single-phase mask with higher diffraction orders [16]. This methodology was used to report POFBGs at the visible region, using a phase mask with a period of 855 nm, allowing to obtain a primary Bragg wavelength at 1,309 nm and a grating with reflection peak

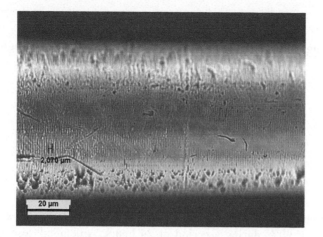

FIGURE 2.33 Microscope image taken at the surface of a PMMA based POF. The fiber has been irradiated through the phase mask method using a phase mask period of 1,033 nm with ±1 principal orders and with 2.4% efficiency on the zeroth diffraction order. The inscription was made with a 248 nm UV radiation with 5 Hz repetition rate and 3 mJ energy during an exposure time of 2 minutes.

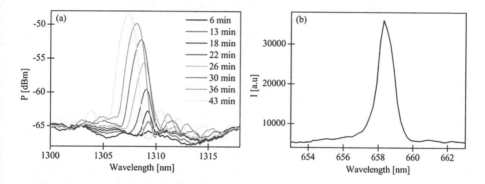

FIGURE 2.34 Reflection spectra of POFBGs obtained, using a high diffraction order phase mask with period of 855 nm, allowing to obtain a primary Bragg wavelength at λ_{Bragg} = 1,309 nm through the ±1 diffraction orders, as seen in (a) and a Bragg wavelength located at $\lambda_{Bragg}/2$ = 659 nm for the ±2 diffraction orders, as shown in (b). (Obtained from G. Statkiewicz-Barabach, et al., "Fabrication of multiple Bragg gratings in microstructured polymer fibers using a phase mask with several diffraction orders," *Opt. Express*, vol. 21, no. 7, pp. 8521–8534, 2013, with permission from the Optical Society of America.)

located at $\lambda_{Bragg}/2$ = 659 nm. The result can be seen in Figure 2.34a, for the Bragg reflection at 1,309 nm for the ±1 diffraction orders (during the inscription) and in Figure 2.34b for the grating written at 659 nm for the ±2 diffraction orders (at the end of the inscription) [16].

Another interesting work regarding POFBGs at the visible region has been made by Woyessa et al. [118], which reported the possibility to shift the resonance Bragg wavelength of a PMMA POFBG written at the 850 nm region down to the visible

TABLE 2.5

Bragg Gratings in POFs of Different Materials, at Short Wavelengths

Material	Type	Dopant	λ_{Bragg} (nm)	1st Report	Reference
TOPAS®	SM-mPOF	-	870	2011	[99]
ZEONEX® 480R	SM-mPOF	-	865	2017	[10]
PC	SM-mPOF	-	892	2017	[122]
CYTOP®	MM-GI	-	605	2018	[123]
PMMA	SM-SI	TSB	843	2018	[124]
PMMA	SM-mPOF	BDK	827	2018	[77]

region (~600 nm), by performing an annealing procedure at a temperature of 80°C and humidity of 90%, during a period of 24 hours. A detailed description of the phenomenon behind the wavelength shift will be described in Subsection 3.2.2. From this particular work, one can point two important advantages, either because it allows the use of less expensive phase masks (i.e. the smaller the pitch the higher the cost), as well as it allows better writing efficiencies, since phase masks at shorter wavelengths become inefficient due to the close proximity between λ_{UV} with Λ_{pm}.

Additional to PMMA, the opportunities of different POF materials and the benefits of working at their low loss region, drove the scientific community to report FBGs in then. Examples of those works may be found in Table 2.5.

2.8 FIBER BRAGG GRATING FABRICATION WITH FEMTOSECOND LASER

Despite the good results achieved for the POFBG inscription using different UV/visible sources in doped/undoped fibers, PMMA is still the most widely available material and the fabrication of Bragg gratings in POFs made of the raw material is much more attractive. Thus, femtosecond lasers are interesting choices for the refractive index modification in those fibers, since they can modify the refractive index using wavelengths that are transparent to the material. This occurs because the strength of the electric field in the focal volume causes nonlinear multiphoton absorption [125].

Pioneer studies on the use of fs lasers for the fabrication of holographic gratings in polymer materials were reported in bulk azodye-doped PMMA [31,32] and in block co-polymers of MMA-EMA, achieving refractive index changes of the order of 3×10^{-3} to 2×10^{-4} [126]. Later, in 2003, Scully et al. reported the use of 800 nm femtosecond laser pulses to write gratings into different undoped PMMA slabs, reaching a maximum refractive index modification of $5 \pm 0.5 \times 10^{-4}$ [125]. The threshold fluence was estimated to be around 30 mJ/cm², while above 800 mJ/cm², machining and damage/filament formation took place, resulting in a decrease of the diffraction efficiency.

In the subsequent years, different works have reported the use of fs laser sources for the inscription of Bragg gratings in POFs. A summary of those, together with their most important parameters may be seen in Table 2.6.

TABLE 2.6
POFBGs Written through the Use of Femtosecond Lasers

Wavelength (nm)	Fiber	Core Material	Method	Year	Reference
387	SM-SI	PMMA/PS	Holographic w/phase mask	2006	[127]
400	SM-SI	PMMA/PS	Phase mask	2008	[22]
800	SM-SI	PMMA/PS	PbP	2009	[128]
800	SM-mPOF	PMMA	PbP	2012	[19]
800	MM-GI	CYTOP®	PbP/LbL	2015	[13]
400	SM-mPOF	PMMA (doped with BDK)	Phase mask	2017	[78]
400	FM-SI	PMMA (doped with TSB & DPS)	Phase mask	2017	[52]
517	MM-SI	Bisphenol-A acrylate	LbL	2018	[129]

2.8.1 PHASE MASK

In 2006, Baum et al. [127] reported the first preliminary study on the inscription of refractive index structures in a PMMA fiber using a frequency doubled femtosecond laser irradiation at 387 nm. The inscription was based on a holographic writing setup in which the diffraction orders of a phase mask (100 lines/mm) were used together with a bi-prism and a focusing microscope objective. The work was performed with a fluence of 120 mJ/cm2 and allowed to visualize refractive index modification using a microscope. However, it was only in 2008 that the first POFBG was demonstrated [22]. The grating was demonstrated at the 1,550 nm region using the phase mask technique and a PMMA SM-POF (core composition: 22 wt% PS and 78 wt% PMMA). The fs laser was a Coherent Mira Rega, operating at 200 kHz, with a wavelength of 800 nm, an output power of 1 W, and pulse duration of 200 fs. The laser was operated at the 2nd harmonic, at the wavelength of 400 nm due to two reasons. First, because higher photon energy allowed to achieve higher refractive index changes compared to the fundamental wavelength, due to the enhanced nonlinear absorption [33]. The second reason was related to the efficiency of the zeroth diffraction order. This occurs because working at the fundamental wavelength (800 nm), the phase mask needs to have a similar period (considering a POFBG at the 1,550 nm region), causing high zeroth diffraction order due to a decrease in diffraction efficiency. As a consequence, the fringe visibility is reduced. Nevertheless, phase masks with zeroth diffraction order lower than 4% can be easily found for wavelengths below 400 nm.

The phase mask setup used by Harbach [22] was similar to the ones implemented for the UV laser systems. The grating was written through a phase mask pitch of 1,040 nm and with a fluence of 0.8 mJ/cm^2. Due to the tiny dimensions of the laser beam at the focus region (1.65 mm in the longitudinal axis of the fiber), the scanning technique with a velocity of 100 mm/min was used, giving an exposure duration of

1 second per scan, with waiting time between scans of ~1–2 minutes. For a total grating length of 5 mm, it took 45 scans, corresponding to 7.2 kJ total dose. The result was a Bragg grating seen in reflection with a bandwidth of 385 pm, refractive index amplitude of 1.2×10^{-4} and negative refractive index change of $~8 \times 10^{-5}$. Despite the interesting results, Harbach also reported that the grating was not stable and vanished after a few days [22]. In an attempt to justify this behavior, Harbach associated the effect to the low fluences used during the irradiation process. However, for higher fluences, the induced damage and melting of the fiber could occur, and thus, the correct optimization of exposure time by proper choice of the scanning speed needs to be properly accessed.

In 2017, Hu et al., motivated to reduce the inscription time of POFBGs decided to implement the same technique used by Harbach. For that they implemented two different SM-PMMA fibers, for mPOF doped with BDK [78] and SI-POF doped with TSB and DPS [52]. The gratings were written through a 1 kHz Ti-sapphire laser and using a phase mask pitch of 1,060 nm. At the laser output, a diaphragm with a diameter of 10 mm was used to select the most uniform part of the beam, followed by a cylindrical lens, to focus the beam on the longitudinal axis of the fiber. The gratings were written with different laser powers and were seen in transmission in real time, allowing to verify their saturation. Regarding the work developed for the mPOF doped with BDK [78], the authors report the capability to write a POFBG centered at 1564.5 nm and with reflectivity of 5, 56, and 83%, for a laser power of 1, 5, and 11 mW and with saturation time of 100, 90, and 40 seconds, respectively. According to the authors, the results were much superior to the ones obtained using the same inscription system in undoped mPOFs, and also using the BDK doped mPOFs and the conventional phase mask technique through the 325 nm HeCd source. In both cases, the gratings required several minutes to be inscribed and also reached lower reflectivity [78]. On what concerns the inscription of the SI-POF doped with TSB and DPS [52], the authors were able to report a Bragg grating written through a 20 mW power in approximately 60 seconds. The grating had a resonance Bragg wavelength centered at 1578.6 nm and a transmission dip of 17 dB, corresponding to a reflectivity of 98%. The refractive index change was negative and its value was about 2×10^{-5}, while the refractive index modulation amplitude reached 1.5×10^{-4}.

Regarding the stability of the gratings, the works developed by Hu et al. [52,78], showed a different behavior than the one achieved by Harbach [22], in which the gratings vanished after a few days. In the works reported by Hu et al., the stability of the gratings was monitored during 1 month for the BDK doped mPOF and fifteen days for the TSB-DPS doped SI-POF. Regarding the previous, a slight decrease of the reflectivity was observed (~5%) and it was associated with the material relaxation due to the release of stress that was frozen after the drawing process and also due to the chemical restructuring [78]. However, for the gratings written in the SI-POF doped with TSB and DPS [52], the authors observed a reflectivity increase for the low reflective gratings (produced with ~15 mW), while a decreased for the high reflective ones (produced with ~20 mW). After this observation, the authors decided to make a post-annealing process, allowing them to achieve gratings with reflectivity similar and higher than the ones recorded right after the inscription process. The result was in agreement with their prior work, regarding the inscription of POFBGs in this fiber with the HeCd laser [72] (see the sub Subsection 2.5.4.1). The reason for such behavior was due to the transition mechanisms between the excited intermediate states of both trans- and cis-isomers and the temperature-dependent glassy polymer matrix, as explained in Ref. [73].

2.8.2   DIRECT WRITING METHODS (POINT-BY-POINT (PBP) AND LINE-BY-LINE (LBL))

One clear advantage of the use of a fs laser is that the geometry where the refractive index needs to be altered, can be precisely controlled by the spot size of the focusing optics and the movement of this spot related to the fiber, either longitudinally (in the direction of guided-light propagation) or transversely (across the core cross-section). While the previous can be used to control the period and phase of the grating, the latter can be used to control the strength of the coupling induced by the grating (i.e., the amplitude of the grating) and in the case of multimode fibers, the differential coupling to the distinct transverse modes [130]. This flexibility in grating design avoids the use of phase masks and thus, it has been adopted for the recording of Bragg gratings in silica fibers [131], as well as in POFs [13,19,128].

The first report of a periodic refractive index modification in a POF through focused femtosecond pulses using the PbP methodology was reported in 2009 by the group of Bang et al. [128]. In this technique, the grating structure is created one point at a time, along the length of the fiber. Each reflector is created separately by a focused laser beam, providing flexibility on the control of the grating period by simply adjusting the position of the focused laser beam along the fiber length. Thus, FBG arrays can be easily implemented in optical fibers by adjusting the period of each grating. A schematic representation of a direct writing femtosecond laser inscription system may be seen in Figure 2.35.

As can be seen from Figure 2.35, the fs writing beam is focused onto the fiber core through an objective lens, allowing to locally change the refractive index. The grating period, defined by the periodicity of the laser-modified volumes, is the ratio of the translation speed of the fiber (in the direction of the guided-light propagation),

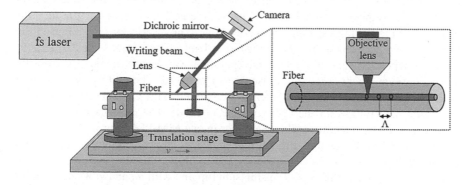

FIGURE 2.35   Representation of a direct writing femtosecond laser inscription system. The translation stage can move freely related to the writing beam, allowing to change the refractive index of the fiber periodically.

to the laser pulse repetition rate. However, when the size of the refractive index modification is higher than the grating period, then, higher-order grating structures need to be designed, allowing to avoid the overlap between two consecutive focusing spots. Thus, it is not surprising to find 2nd, 3rd, and 4th order gratings with this technique.

According to the work developed in Ref. [128], the threshold pulse energy for damage in a POF using a 1 kHz, 800 nm fs laser with 120 s pulse duration, is approximately 60 nJ, allowing to create refractive index modifications in the fiber core. However, results from this work were only observed through microscope images. Stefani et al. in 2012 [19], used the same technology in a PMMA fiber made of a special microstructured hole arrangement. The grating was fabricated using a femtosecond laser system with 100 fs pulses and with a central wavelength of 800 nm, a repetition rate of 1 kHz and pulse energy of 75 nJ, giving a fluence of ~1 J/cm^2. In order to have a single pulse per spot (without overlap between consecutive spots) and to have a grating located at around 1,520 nm (considering an effective refractive index of 1.4804), the authors implemented a 4th order grating, giving a translation stage velocity of 2.053 mm/s. The grating length was chosen to be equal to 5 mm, which required only 2.5 seconds of writing time [19].

Since the refractive index modification in these types of gratings does not rely on the photosensitivity of the polymer material, the authors suggested that these type of gratings could be interesting for perfluorinated fibers, which are known to be the most transparent fiber for the wavelengths between the 800 nm and 1,300 nm (see Figure 1.4). The demonstration of gratings in these type of fibers, was shown in 2015 by Kalli et al. [13], using either the PbP and LbL methods through a 800 nm fs laser, in a commercial MM GI-POF (fiber composed of 62.5 μm perfluorinated GI core and 500 μm protective overcladding, made of PC and PS). One clear advantage of the use of 800 nm fs laser over the UV inscription methods [12,124] is that the protective overcladding does not need to be removed for the inscription process, making it much more attractive when compared with UV inscription methods, which require the etching of the fiber to avoid excessive attenuation by the protective overcladding [12,123,132]. In both PbP and LbL methods, the longitudinal position of the modifications determines the period and phase, however, in the LbL method, transverse movement is also allowed, offering greater flexibility in the grating properties, since the grating's spatial size is controlled. The inscription of gratings of this type has been performed through the use of a femtosecond laser with 220 fs pulse duration, repetition rate of 1 KHz and operation wavelength of 517 nm from the second harmonic conversion and pulse energy of 80 nJ. The gratings were written with a length of 3.2 mm and were intended to operate at the C-band region using 4th and 2nd order gratings for the LbL and PbP methods. The gratings were observed in transmission by injecting a silica SM fiber in one MM-POF terminal and by coupling the other POF terminal directly into an OSA. The launching conditions were optimized to excite the low order modes and by doing that, the authors were able to report a POFBG with 5.5 dB transmission dip and with a bandwidth of 0.25 nm. The number of fiber modes coupling to the grating due to the MM nature of the fiber (capability to propagate ~140 modes at the C-band) generated a multi-peak behavior. Because of that, in a subsequent publication from the same authors, they reported the capability to reduce the number of fiber modes coupling to the grating, by limiting the LbL grating to the central part of the core, where the gradient index profile peaks [133]. The tests were made for grating planes with widths of 5, 10, and 30 μm and grating planes between 300 and 1,000 planes, allowing to conclude that the best results (cleanest spectrum showing preferable excitation to the strongest lower order modes), could be achieved for gratings presenting plane widths lower than 15 μm and with the number of grating planes between 300 and 500 [133].

The flexibility in changing the writing parameters such as the grating period along the grating length; or the distance between two consecutive Bragg gratings; and also the capability to preferentially excite the strongest higher-order modes, allowed the group of Kalli et al. to report FBG arrays [133,134], chirped POFBGs [133] and also FP cavities [133], being most of those gratings used for sensing applications.

REFERENCES

1. K. O. Hill, Y. Fujii, D. C. Johnson, and B. S. Kawasaki, "Photosensitivity in optical fiber waveguides: Application to reflection filter fabrication," *Appl. Phys. Lett.*, vol. 32, no. 10, pp. 647–649, 1978.
2. A. Othonos, "Fiber Bragg gratings," *Rev. Sci. Instrum.*, vol. 68, no. 12, pp. 4309–4341, 1997.
3. G. D. Peng, Z. Xiong, and P. L. Chu, "Photosensitivity and gratings in dye-doped polymer optical fibers," *Opt. Fiber Technol.*, vol. 5, no. 2, pp. 242–251, 1999.
4. Z. Xiong, G. D. Peng, B. Wu, and P. L. Chu, "Highly tunable Bragg gratings in single-mode polymer optical fibers," *IEEE Photonics Technol. Lett.*, vol. 11, no. 3, pp. 352–354, 1999.
5. M. van Eijkelenborg et al., "Microstructured polymer optical fibre," *Opt. Express*, vol. 9, no. 7, pp. 319–327, 2001.
6. H. Dobb, D. J. Webb, K. Kalli, A. Argyros, M. C. J. Large, and M. A. van Eijkelenborg, "Continuous wave ultraviolet light-induced fiber Bragg gratings in few - and single-mode microstructured polymer optical fibers," *Opt. Lett.*, vol. 30, no. 24, pp. 3296–3298, 2005.
7. I. P. Johnson et al., "Optical fibre Bragg grating recorded in TOPAS cyclic olefin copolymer," *Electron. Lett.*, vol. 47, no. 4, pp. 271–272, 2011.
8. A. Fasano et al., "Fabrication and characterization of polycarbonate microstructured polymer optical fibers for high-temperature-resistant fiber Bragg grating strain sensors," *Opt. Mater. Express*, vol. 6, no. 2, pp. 649–659, 2016.
9. G. Statkiewicz-Barabach, D. Kowal, P. Mergo, and W. Urbanczyk, "Comparison of growth dynamics and temporal stability of Bragg gratings written in polymer fibers of different types," *J. Opt.*, vol. 17, no. 8, p. 085606, 2015.
10. G. Woyessa, A. Fasano, C. Markos, A. Stefani, H. K. Rasmussen, and O. Bang, "Zeonex microstructured polymer optical fiber: Fabrication friendly fibers for high temperature and humidity insensitive Bragg grating sensing," *Opt. Mater. Express*, vol. 7, no. 1, pp. 286–295, 2017.
11. R. Oliveira, T. H. R. Marques, L. Bilro, R. Nogueira, and C. M. B. Cordeiro, "Multiparameter POF sensing based on multimode interference and fiber Bragg grating," *J. Light. Technol.*, vol. 35, no. 1, pp. 3–9, 2017.
12. M. Koerdt et al., "Fabrication and characterization of Bragg gratings in perfluorinated polymer optical fibers and their embedding in composites," *Mechatronics*, vol. 34, pp. 137–146, 2016.

13. A. Lacraz, M. Polis, A. Theodosiou, C. Koutsides, and K. Kalli, "Femtosecond laser inscribed Bragg gratings in low loss CYTOP polymer optical fiber," *IEEE Photonics Technol. Lett.*, vol. 27, no. 7, pp. 693–696, 2015.

14. C. A. F. Marques, L. Bilro, N. J. Alberto, D. J. Webb, and R. Nogueira, "Narrow bandwidth Bragg gratings imprinted in polymer optical fibers for different spectral windows," *Opt. Commun.*, vol. 307, pp. 57–61, 2013.

15. I.-L. Bundalo, K. Nielsen, C. Markos, and O. Bang, "Bragg grating writing in PMMA microstructured polymer optical fibers in less than 7 minutes," *Opt. Express*, vol. 22, no. 5, pp. 5270–5276, 2014.

16. G. Statkiewicz-Barabach, K. Tarnowski, D. Kowal, P. Mergo, and W. Urbanczyk, "Fabrication of multiple Bragg gratings in microstructured polymer fibers using a phase mask with several diffraction orders," *Opt. Express*, vol. 21, no. 7, pp. 8521–8534, 2013.

17. H. Y. Liu, H. B. Liu, G. D. Peng, and P. L. Chu, "Observation of type I and type II gratings behavior in polymer optical fiber," *Opt. Commun.*, vol. 220, no. 4–6, pp. 337–343, 2003.

18. H. B. Liu, H. Y. Liu, G. D. Peng, and P. L. Chu, "Novel growth behaviors of fiber Bragg gratings in polymer optical fiber under UV irradiation with low power," *IEEE Photonics Technol. Lett.*, vol. 16, no. 1, pp. 159–161, 2004.

19. A. Stefani, M. Stecher, G. E. Town, and O. Bang, "Direct writing of fiber Bragg grating in microstructured polymer optical fiber," *IEEE Photonics Technol. Lett.*, vol. 24, no. 13, pp. 1148–1150, 2012.

20. W. J. Tomlinson, I. P. Kaminow, E. A. Chandross, R. L. Fork, and W. T. Silfvast, "Photoinduced refractive index increase in poly(methylmethacrylate) and its applications," *Appl. Phys. Lett.*, vol. 16, no. 12, pp. 486–489, 1970.

21. M. Bowden, E. Chandross, and I. Kaminow, "Mechanism of the photoinduced refractive index increase in polymethyl methacrylate," *Appl. Opt.*, vol. 13, no. 1, pp. 112–117, 1974.

22. N. Harbach, *Fiber Bragg Gratings in Polymer Optical Fibers*. Lausanne: École Polytechnique Fédérale de Lausanne, 2008.

23. M. Kopietz, M. Lechner, and D. Steinmeier, "Light-induced refractive index changes in polymethylmethacrylate (PMMA) blocks," *Polym. Photochem.*, vol. 5, pp. 109–119, 1984.

24. A. Torikai, M. Ohno, and K. Fueki, "Photodegradation of poly(methyl methacrylate) by monochromatic light - Quantum yield, effect of wavelengths, and light-intensity," *J. Appl. Polym. Sci.*, vol. 41, no. 5–6, pp. 1023–1032, 1990.

25. T. Mitsuoka, A. Torikai, and K. Fueki, "Wavelength sensitivity of the photodegradation of poly (methyl methacrylate)," *J. Appl. Polym. Sci.*, vol. 47, pp. 1027–1032, 1993.

26. S. Küper and M. Stuke, "UV-excimer-laser ablation of polymethylmethacrylate at 248 nm: Characterization of incubation sites with Fourier transform IR- and UV-Spectroscopy," *Appl. Phys. A Solids Surfaces*, vol. 49, no. 2, pp. 211–215, 1989.

27. R. Srinivasan, B. Braren, R. W. Dreyfus, L. Hadel, and D. E. Seeger, "Mechanism of the ultraviolet laser ablation of polymethyl methacrylate at 193 and 248 nm: Laser-induced fluorescence analysis, chemical analysis, and doping studies," *J. Opt. Soc. Am. B*, vol. 3, no. 5, pp. 785–791, 1986.

28. R. Srinivasan, B. Braren, and K. G. Casey, "Nature of 'incubation pulses' in the ultraviolet laser ablation of polymethyl methacrylate," *J. Appl. Phys.*, vol. 68, no. 4, pp. 1842–1847, 1990.

29. A. K. Baker and P. E. Dyer, "Refractive-index modification of polymethylmethacrylate (PMMA) thin films by KrF-laser irradiation," *Appl. Phys. A Solids Surfaces*, vol. 57, no. 6, pp. 543–544, 1993.

30. C. Wochnowski, S. Metev, and G. Sepold, "UV–laser-assisted modification of the optical properties of polymethylmethacrylate," *Appl. Surf. Sci.*, vol. 154, pp. 706–711, 2000.

31. J. Zhai, Y. Shen, J. Si, J. Qiu, and K. Hirao, "The fabrication of permanent holographic gratings in bulk polymer medium by a femtosecond laser," *J. Phys. D. Appl. Phys.*, vol. 34, no. 24, pp. 3466–3469, 2001.

32. J. Si, J. Qiu, J. Zhai, Y. Shen, and K. Hirao, "Photoinduced permanent gratings inside bulk azodye-doped polymers by the coherent field of a femtosecond laser," *Appl. Phys. Lett.*, vol. 80, no. 3, pp. 359–361, 2002.

33. A. Baum et al., "Photochemistry of refractive index structures in poly (methyl methacrylate) by femtosecond laser irradiation," *Opt. Lett.*, vol. 32, no. 2, pp. 190–192, 2007.

34. T. Erdogan, "Fiber grating spectra," *J. Light. Technol.*, vol. 15, no. 8, pp. 1277–1294, 1997.

35. G. Meltz, W. W. Morey, and W. H. Glenn, "Formation of Bragg gratings in optical fibers by a transverse holographic method," *Opt. Lett.*, vol. 14, no. 15, pp. 823–825, 1989.

36. A. Cusano, A. Cutolo, and J. Albert, *Fiber Bragg Grating Sensors: Research Advancements, Industrial Applications and Market Exploitation*, 1st ed. Sharjah, United Arab Emirates, Bentham Science Publishers Ltd., 2011.

37. R. Kashyap, *Fiber Bragg Gratings*. San Diego, CA, Elsevier, 2010.

38. K. O. Hill, B. Malo, F. Bilodeau, D. C. Johnson, and J. Albert, "Bragg gratings fabricated in monomode photosensitive optical fiber by UV exposure through a phase mask," *Appl. Phys. Lett.*, vol. 62, no. 10, pp. 1035–1037, 1993.

39. P. L. Chu and G.-D. Peng, "Photosensitivities in germanium-doped planar waveguides and dye-doped polymer optical fibres," in *Photorefractive Fiber and Crystal Devices: Materials, Optical Properties, and Applications IV*, F. T. S. Yu and S. Yin, Eds. Bellingham: SPIE, 1998, vol. 3470, pp. 120–127.

40. G. D. Peng, P. L. Chu, Z. Xiong, T. W. Whitbread, and R. P. Chaplin, "Dye-doped step-index polymer optical fiber for broadband optical amplification," *J. Light. Technol.*, vol. 14, no. 10, pp. 2215–2223, 1996.

41. H. Liu, G. Peng, and P. Chu, "Polymer fiber Bragg gratings with 28-dB transmission rejection," *IEEE Photonics Technol. Lett.*, vol. 14, no. 7, pp. 935–937, 2002.

42. J. Yu, X. Tao, and H. Tam, "Trans-4-stilbenemethanol-doped photosensitive polymer fibers and gratings," *Opt. Lett.*, vol. 29, no. 2, pp. 156–158, 2004.

43. Z. Xiong, G. D. Peng, B. Wu, and P. L. Chu, "Effects of the zeroth-order diffraction of a phase mask on Bragg gratings," *J. Light. Technol.*, vol. 17, no. 11, pp. 2361–2365, 1999.

44. G. Rajan, M. Y. Mohd Noor, N. H. Lovell, E. Ambikaizrajah, G. Farrell, and G.-D. Peng, "Polymer micro-fiber Bragg grating," *Opt. Lett.*, vol. 38, no. 17, pp. 3359–3362, 2013.

45. Z. F. Zhang, C. Zhang, X. M. Tao, G. F. Wang, and G. D. Peng, "Inscription of polymer optical fiber Bragg grating at 962 nm and its potential in strain sensing," *IEEE Photonics Technol. Lett.*, vol 22, no. 21, pp. 1562–1564, 2010.

46. X. Hu, C.-F. J. Pun, H.-Y. Tam, P. Mégret, and C. Caucheteur, "Highly reflective Bragg gratings in slightly etched step-index polymer optical fiber," *Opt. Express*, vol. 22, no. 15, pp. 18807–18817, 2014.

47. R. Kashyap, *Fiber Bragg Gratings*. San Diego, CA: Academic Press, 1999.

48. J.-L. Archambault, L. Reekie, and P. St. J. Russell, "100% reflectivity Bragg reflectors produced in optical fibres by single excimer laser pulses," *Electron. Lett.*, vol. 29, no. 5, pp. 453–455, 1993.

49. W. Yuan et al., "Improved thermal and strain performance of annealed polymer optical fiber Bragg gratings," *Opt. Commun.*, vol. 284, no. 1, pp. 176–182, 2011.

50. R. Oliveira, L. Bilro, and R. Nogueira, "Bragg gratings in a few mode microstructured polymer optical fiber in less than 30 seconds," *Opt. Express*, vol. 23, no. 8, pp. 10181–10187, 2015.

51. G. Rajan, B. Liu, Y. Luo, E. Ambikairajah, and G.-D. Peng, "High sensitivity force and pressure measurements using etched singlemode polymer fiber Bragg gratings," *IEEE Sens. J.*, vol. 13, no. 5, pp. 1794–1800, 2013.

52. X. Hu, D. Kinet, K. Chah, C.-F. J. Pun, H.-Y. Tam, and C. Caucheteur, "Bragg grating inscription in PMMA optical fibers using 400-nm femtosecond pulses," *Opt. Lett.*, vol. 42, no. 14, pp. 2794–2797, 2017.

53. K. Bhowmik et al., "Etching process related changes and effects on solid-core single-mode polymer optical fiber grating," *IEEE Photonics J.*, vol. 8, no. 1, p. 2500109, 2016.

54. A. Stefani, C. Markos, and O. Bang, "Narrow bandwidth 850-nm fiber Bragg gratings in few-mode polymer optical fibers," *IEEE Photonics Technol. Lett.*, vol. 23, no. 10, pp. 660–662, 2011.

55. J. Terblanche, D. Schmieder, and J. Meyer, "Fibre Bragg gratings in polymer optical fibres at 980 nm," in *20th International Conference on Optical Fibre Sensors*, J. Jones et al., Eds. Bellingham: SPIE, 2009, vol. 7503, p. 75037F.

56. Y. Luo, Q. Zhang, H. Liu, and G.-D. Peng, "Gratings fabrication in benzildimethylketal doped photosensitive polymer optical fibers using 355 nm nanosecond pulsed laser," *Opt. Lett.*, vol. 35, no. 5, pp. 751–753, 2010.

57. Z. Li, H. Y. Tam, L. Xu, and Q. Zhang, "Fabrication of long-period gratings in poly(methyl methacrylate-co-methyl vinyl ketone-co-benzyl methacrylate)-core polymer optical fiber by use of a mercury lamp," *Opt. Lett.*, vol. 30, no. 10, pp. 1117–1119, 2005.

58. T. Wang et al., "Enhancing photosensitivity in near UV/vis band by doping 9-vinylanthracene in polymer optical fiber," *Opt. Commun.*, vol. 307, pp. 5–8, 2013.

59. J. M. Yu, X. M. Tao, and H. Y. Tam, "Fabrication of UV sensitive single-mode polymeric optical fiber," *Opt. Mater.*, vol. 28, no. 3, pp. 181–188, 2006.

60. J. Bonefacino et al., "Ultra-fast polymer optical fibre Bragg grating inscription for medical devices," *Light Sci. Appl.*, vol. 7, no. 3, p. 17161, 2018.

61. Y. Luo et al., "Birefringent azopolymer long period fiber gratings induced by 532 nm polarized laser," *Opt. Commun.*, vol. 282, no. 12, pp. 2348–2353, 2009.

62. Y. Luo et al., "Optical manipulable polymer optical fiber Bragg gratings with azopolymer as core material," *Appl. Phys. Lett.*, vol. 91, no. 2007, pp. 2005–2008, 2007.

63. X. Xingsheng et al., "Birefringent gratings induced by polarized laser in azobenzene-doped Poly(methyl methecrylate) optical fibers," in *Fiber Optics and Optoelectronics for Network Applications*, J. Liu and Z. Wang, Eds. Bellingham: SPIE, 2001, vol. 4603, pp. 260–265.

64. R. Van Boxel, *Bragg Gratings in Photosensitive Graded Index Polymer Optical Fibres.* Leuven: Katholieke Universiteit Leuven, 2005.

65. O. H. Park, J. I. Jung, and B. S. Bae, "Photoinduced condensation of sol-gel hybrid glass films doped with benzildimethylketal," *J. Mater. Res.*, vol. 16, no. 7, pp. 2143–2148, 2001.

66. Y. Luo, B. Yan, Q. Zhang, G.-D. Peng, J. Wen, and J. Zhang, "Fabrication of polymer optical fibre (POF) gratings," *Sensors*, vol. 17, no. 511, 2017.

67. D. R. Tyler, "Mechanistic aspects of the effects of stress on the rates of photochemical degradation reactions in polymers," *J. Macromol. Sci. - Polym. Rev.*, vol. 44, no. 4, pp. 351–388, 2004.

68. D. Sáez-Rodríguez, K. Nielsen, O. Bang, and D. J. Webb, "Photosensitivity mechanism of undoped poly(methyl methacrylate) under UV radiation at 325 nm and its spatial resolution limit," *Opt. Lett.*, vol. 39, no. 12, pp. 3421–3424, 2014.

69. X. M. Tao, J. M. Yu, and H.-Y. Tam, "Photosensitive polymer optical fibres and gratings," *Trans. Inst. Meas. Control*, vol. 29, no. 3–4, pp. 255–270, 2007.

70. X. Tao, J. Yu, M. Suleyman Demokan, H. Tam, and D. Yang, "Photosensitivity and grating development in trans-4-stilbenemethanol-doped poly(methyl methacrylate) materials," *Opt. Commun.*, vol. 265, pp. 132–139, 2006.

71. D. Sáez-Rodríguez, K. Nielsen, O. Bang, and D. J. Webb, "Time-dependent variation of fibre Bragg grating reflectivity in PMMA based polymer optical fibres," *Opt. Lett.*, vol. 40, no. 7, pp. 1476–1479, 2015.

72. X. Hu, D. Kinet, P. Mégret, and C. Caucheteur, "Control over photo-inscription and thermal annealing to obtain high-quality Bragg gratings in doped PMMA optical fibers," *Opt. Lett.*, vol. 41, no. 13, pp. 2930–2933, 2016.
73. S. Malkin and E. Fischer, "Temperature dependence of photoisomerization. III. Direct and sensitized photoisomerization of stilbenes," *J. Phys. Chem.*, vol. 68, no. 5, pp. 1153–1163, 1964.
74. H. Franke, "Optical recording of refractive-index patterns in doped poly-(methyl methacrylate) films," *Appl. Opt.*, vol. 23, no. 16, pp. 2729–2733, 1984.
75. X. Cheng et al., "High-sensitivity temperature sensor based on Bragg grating in BDK-doped photosensitive polymer optical fiber," *Chinese Opt. Lett.*, vol. 9, no. 2, p. 020602, 2011.
76. X. Cheng, J. Bonefacino, B. O. Juan, and H. Y. Tam, "All-polymer fiber-optic pH sensor," *Opt. Express*, vol. 26, no. 11, pp. 14610–14616, 2018.
77. D. Sáez-Rodríguez, K. Nielsen, H. K. Rasmussen, O. Bang, and D. J. Webb, "Highly photosensitive polymethyl methacrylate microstructured polymer optical fiber with doped core," *Opt. Lett.*, vol. 38, no. 19, pp. 3769–3772, 2013.
78. X. Hu et al., "BDK-doped core microstructured PMMA optical fiber for effective Bragg grating photo-inscription," *Opt. Lett.*, vol. 42, no. 11, pp. 2209–2212, 2017.
79. A. Pospori, C. A. F. Marques, O. Bang, D. J. Webb, and P. André, "Polymer optical fiber Bragg grating inscription with a single UV laser pulse," *Opt. Express*, vol. 25, no. 8, pp. 9028–9038, 2017.
80. L. Pereira et al., "Polymer optical fiber Bragg grating inscription with a single Nd: YAG laser pulse," *Opt. Express*, vol. 26, no. 14, pp. 18096–18104, 2018.
81. G. Rajan, K. Bhowmik, J. Xi, and G. D. Peng, "Etched polymer fibre bragg gratings and their biomedical sensing applications," *Sensors*, vol. 17, no. 10, p. 2336, 2017.
82. G. Rajan and G.-D. Peng, "Fabrication and characterization of a polymer micro-fiber Bragg grating," in *Fourth Asia Pacific Optical Sensors Conference*, M. Yang, D. Wang and Y. Rao, Eds. Bellingham: SPIE, 2013, vol. 8924, p. 892432.
83. M. C. J. Large, L. Poladian, G. W. Barton, and M. A. van Eijkelenborg, *Microstructured Polymer Optical Fibres*. Boston, MA: Springer US, 2008.
84. J. C. Knight, T. A. Birks, P. S. J. Russell, and D. M. Atkin, "All-silica single-mode optical fiber with photonic crystal cladding," *Opt. Lett.*, vol. 21, no. 19, pp. 1547–1549, 1996.
85. T. A. Birks, J. C. Knight, and P. S. Russell, "Endlessly single-mode photonic crystal fiber," *Opt. Lett.*, vol. 22, no. 13, pp. 961–963, 1997.
86. R. Oliveira et al., "Bragg gratings inscription in highly birefringent microstructured POFs," *IEEE Photonics Technol. Lett.*, vol. 28, no. 6, pp. 621–624, 2016.
87. N. A. Issa, M. A. van Eijkelenborg, M. Fellew, F. Cox, G. Henry, and M. C. J. Large, "Fabrication and study of microstructured optical fibers with elliptical holes," *Opt. Lett.*, vol. 29, no. 12, pp. 1336–1338, 2004.
88. M. K. Szczurowski, T. Martynkien, G. Statkiewicz-Barabach, W. Urbanczyk, and D. J. Webb, "Measurements of polarimetric sensitivity to hydrostatic pressure, strain and temperature in birefringent dual-core microstructured polymer fiber," *Opt. Express*, vol. 18, no. 12, pp. 12076–12087, 2010.
89. T. Martynkien, P. Mergo, and W. Urbanczyk, "Sensitivity of birefringent microstructured polymer optical fiber to hydrostatic pressure," *IEEE Photonics Technol. Lett.*, vol. 25, no. 16, pp. 1562–1565, 2013.
90. J. Olszewski, P. Mergo, K. Gąsior, and W. Urbańczyk, "Highly birefringent microstructured polymer fibers optimized for a preform drilling fabrication method," *J. Opt.*, vol. 15, no. 075713, 2013.
91. D. Wang and L. Wang, "Design of Topas microstructured fiber with ultra-flattened chromatic dispersion and high birefringence," *Opt. Commun.*, vol. 284, pp. 5568–5571, 2011.

92. Y. Zhang et al., "Casting preforms for microstructured polymer optical fibre fabrication," *Opt. Express*, vol. 14, no. 12, pp. 5541–5547, 2006.

93. K. Cook et al., "Air-structured optical fiber drawn from a 3D-printed preform," *Opt. Lett.*, vol. 40, no. 17, pp. 3966–3969, 2015.

94. M. A. van Eijkelenborg et al., "Recent progress in microstructured polymer optical fibre fabrication and characterisation," *Opt. Fiber Technol.*, vol. 9, no. 4, pp. 199–209, 2003.

95. M. Large, S. Ponrathnam, A. Argyros, N. Pujari, and F. Cox, "Solution doping of microstructured polymer optical fibres," *Opt. Express*, vol. 12, no. 9, pp. 1966–1971, 2004.

96. D. Kowal, G. Statkiewicz-Barabach, P. Mergo, and W. Urbanczyk, "Inscription of long period gratings using an ultraviolet laser beam in the diffusion-doped microstructured polymer optical fiber," *Appl. Opt.*, vol. 54, no. 20, pp. 6327–6333, 2015.

97. K. E. Carroll, C. Zhang, D. J. Webb, K. Kalli, A. Argyros, and M. C. Large, "Thermal response of Bragg gratings in PMMA microstructured optical fibers," *Opt. Express*, vol. 15, no. 14, pp. 8844–8850, 2007.

98. D. J. Webb et al., "Recent developments of Bragg gratings in PMMA and TOPAS polymer optical fibers," in *Advanced Sensor Systems and Applications III*, Y.-J. Rao, Y. Liao and G.-D. Peng, Eds. Bellingham: SPIE, 2007, vol. 6830, p. 683002.

99. W. Yuan et al., "Humidity insensitive TOPAS polymer fiber Bragg grating sensor," *Opt. Express*, vol. 19, no. 20, pp. 19731–19739, 2011.

100. I. P. Johnson, K. Kalli, and D. J. Webb, "827 nm Bragg grating sensor in multimode microstructured polymer optical fibre," *Electron. Lett.*, vol. 46, no. 17, pp. 1217–1218, 2010.

101. G. D. Marshall, D. J. Kan, A. A. Asatryan, L. C. Botten, and M. J. Withford, "Transverse coupling to the core of a photonic crystal fiber: the photo-inscription of gratings," *Opt. Express*, vol. 15, no. 12, pp. 7876–7887, 2007.

102. R. Oliveira, L. Bilro, J. Heidarialamdarloo, and R. Nogueira, "Fast inscription of Bragg grating arrays in undoped PMMA mPOF," in *24th International Conference on Optical Fiber Sensors*, 2015, vol. 9634, p. 96344X.

103. I.-L. Bundalo, K. Nielsen, and O. Bang, "Angle dependent fiber Bragg grating inscription in microstructured polymer optical fibers," *Opt. Express*, vol. 23, no. 3, pp. 3699–3707, 2015.

104. K.-C. D. Cheng et al., "Optimization of 3-hole-assisted PMMA optical fiber with double cladding for UV-induced FBG fabrication," *Opt. Express*, vol. 17, no. 4, pp. 2080–2088, 2009.

105. W. Wu et al., "Design and fabrication of single mode polymer optical fiber gratings," *J. Optoelectron. Adv. Mater.*, vol. 12, no. 8, pp. 1652–1659, 2010.

106. H. Dobb et al., "Grating based devices in polymer optical fibre," in *Optical Sensing II*, B. Culshaw, A. G. Mignani, H. B. Bartelt and L. R. Jaroszewicz, Eds. Bellingham: SPIE, 2006, vol. 6189, p. 618901.

107. K. Kalli et al., "Non-linear temperature response of Bragg gratings in doped and un-doped holey polymer optical fibre," in *Photonic Crystal Fibers*, K. Kalli, Ed. Bellingham: SPIE 2007, vol. 6588, p. 65880E.

108. H. Y. Liu, G. D. Peng, and P. L. Chu, "Thermal tuning of polymer optical fiber Bragg gratings," *IEEE Photonics Technol. Lett.*, vol. 13, no. 8, pp. 824–826, 2001.

109. G. N. Harbach, H. G. Limberger, and R. P. Salathé, "Influence of humidity and temperature on polymer optical fiber Bragg gratings," in *Advanced Photonics & Renewable Energy*, 2010, p. BTuB2.

110. F. C. Burns and S. R. Cain, "The effect of pulse repetition rate on laser ablation of polyimide and polymers," *J. Phys. D. Appl. Phys.*, vol. 29, no. 5, pp. 1349–1355, 1996.

111. T. Ishigure, E. Nihei, and Y. Koike, "Optimum refractive-index profile of the graded-index polymer optical fiber, toward gigabit data links," *Appl. Opt.*, vol. 35, no. 12, pp. 2048–2053, 1996.

112. R. Nogueira, R. Oliveira, L. Bilro, and J. Heidarialamdarloo, "New advances in polymer fiber Bragg gratings," *Opt. Laser Technol.*, vol. 78, pp. 104–109, 2016.
113. R. Oliveira, L. Bilro, J. Heidarialamdarloo, and R. Nogueira, "Fabrication and characterization of polymer fiber Bragg gratings inscribed with KrF UV laser," in *26th International Conference on Plastic Optical Fibres*, 2015, pp. 371–375.
114. C. Markos, A. Stefani, K. Nielsen, H. K. Rasmussen, W. Yuan, and O. Bang, "High-Tg TOPAS microstructured polymer optical fiber for fiber Bragg grating strain sensing at 110 degrees," *Opt. Express*, vol. 21, no. 4, pp. 4758–4765, 2013.
115. G. Woyessa et al., "Zeonex-PMMA microstructured polymer optical FBGs for simultaneous humidity and temperature sensing," *Opt. Lett.*, vol. 42, no. 6, pp. 1161–1164, 2017.
116. D. J. Webb et al., "Temperature sensitivity of Bragg gratings in PMMA and TOPAS microstructured polymer optical fibres," in *Photonic Crystal Fibers II*, K. Kalli and W. Urbanczyk, Eds. Bellingham: SPIE, 2008, vol. 6990, p. 69900L.
117. C. A. F. Marques et al., "Fast and stable gratings inscription in POFs made of different materials with pulsed 248 nm KrF laser," *Opt. Express*, vol. 26, no. 2, pp. 2013–2022, 2018.
118. G. Woyessa, K. Nielsen, A. Stefani, C. Markos, and O. Bang, "Temperature insensitive hysteresis free highly sensitive polymer optical fiber Bragg grating humidity sensor," *Opt. Express*, vol. 24, no. 2, pp. 1206–1213, 2016.
119. Z. F. Zhang and X. Ming Tao, "Synergetic effects of humidity and temperature on PMMA based fiber Bragg gratings," *J. Light. Technol.*, vol. 30, no. 6, pp. 841–845, 2012.
120. I. P. Kaminow, H. P. Weber, and E. A. Chandross, "Poly(methyl methacrylate) dye laser with internal diffraction grating resonator," *Appl. Phys. Lett.*, vol. 18, no. 11, pp. 497–499, 1971.
121. C. M. Rollinson, S. A. Wade, B. P. Kouskousis, D. J. Kitcher, G. W. Baxter, and S. F. Collins, "Variations of the growth of harmonic reflections in fiber Bragg gratings fabricated using phase masks," *J. Opt. Soc. Am. A*, vol. 29, no. 7, pp. 1259–1268, 2012.
122. G. Woyessa, A. Fasano, C. Markos, H. Rasmussen, and O. Bang, "Low loss polycarbonate polymer optical fiber for high temperature FBG humidity sensing," *IEEE Photonics Technol. Lett.*, vol. 29, no. 7, pp. 575–578, 2017.
123. R. Min, B. Ortega, A. Leal-Junior, and C. Marques, "Fabrication and characterization of Bragg grating in CYTOP POF at 600-nm wavelength," *IEEE Sensors Lett.*, vol. 2, no. 3, 2018.
124. R. Min et al., "Bragg gratings inscription in TS-doped PMMA POF by using 248-nm KrF pulses," *IEEE Photonics Technol. Lett.*, vol. 30, no. 18, pp. 1609–1612, 2018.
125. P. J. Scully, D. Jones, and D. A. Jaroszynski, "Femtosecond laser irradiation of polymethylmethacrylate for refractive index gratings," *J. Opt. A Pure Appl. Opt.*, vol. 5, no. 4, pp. S92–S96, 2003.
126. S. Katayama, M. Horiike, K. Hirao, and N. Tsutsumi, "Diffraction measurement of grating structure induced by irradiation of femtosecond laser pulse in acrylate block copolymers," *Jpn. J. Appl. Phys.*, vol. 41, no. 4A, pp. 2155–2162, 2002.
127. A. Baum, P. J. Scully, and W. Perrie, "Femtosecond laser modification of poly(methyl methacrylate) at 387 nm wavelength," in *2006 Conference on Lasers and Electro-Optics (CLEO)*, 2006, p. JTuD12.
128. M. Stecher et al., "Periodic refractive index modifications inscribed in polymer optical fibre by focussed femtosecond pulses," in *18th International Conference on Plastic Optical Fibers*, 2009.
129. A. Leal-Junior et al., "Characterization of a new polymer optical fiber with enhanced sensing capabilities using a Bragg grating," *Opt. Lett.*, vol. 43, no. 19, pp. 4799–4802, 2018.

130. G. D. Marshall, R. J. Williams, N. Jovanovic, M. J. Steel, and M. J. Withford, "Point-by-point written fiber-Bragg gratings and their application in complex grating designs," *Opt. Express*, vol. 18, no. 19, pp. 19844–19859, 2010.

131. E. Wikszak, J. Burghoff, M. Will, S. Nolte, A. Tunnermann, and T. Gabler, "Recording of fiber Bragg gratings with femtosecond pulses using a 'point by point' technique," in *Conference on Lasers and Electro-Optics (CLEO)*, 2004, p. CThM7.

132. Y. Zheng, K. Bremer, and B. Roth, "Investigating the strain, temperature and humidity sensitivity of a multimode graded-index perfluorinated polymer optical fiber with Bragg grating," *Sensors*, vol. 18, no. 5, p. 1436, 2018.

133. A. Theodosiou, A. Lacraz, A. Stassis, M. Komodromos, and K. Kalli, "Plane-by-Plane femtosecond laser inscription method for single-peak Bragg gratings in multimode CYTOP polymer optical fibre," *J. Light. Technol.*, vol. 35, no. 24, pp. 5404–5410, 2017.

134. A. Lacraz, A. Theodosiou, and K. Kalli, "Femtosecond laser inscribed Bragg grating arrays in long lengths of polymer optical fibres; a route to practical sensing with POF," *Electron. Lett.*, vol. 52, no. 19, pp. 1626–1627, 2016.

3 Special Polymer Optical Fiber Bragg Gratings

3.1 INTRODUCTION

In recent years the technology on the fabrication of fiber Bragg gratings (FBGs) in polymer optical fibers (POFs) has been intensively investigated, allowing the fabrication of polymer optical fiber Bragg gratings (POFBGs) with different properties. Recent progress on this fiber technology allowed the creation of POFBGs in different fiber structures, such as step-index (SI) [1], no-core [2], microstructured [3], graded-index (GI) [4,5], highly-birefringent (HiBi) [6], etc. The wide availability of polymers makes them attractive to explore new fiber sensing technologies. FBGs in POFs based on undoped materials, such as polymethylmethacrylate (PMMA) [1,3], TOPAS® [7], Polycarbonate (PC) [8], ZEONEX® [2,9], and CYTOP® [4,5], have already been demonstrated, allowing the development of fiber sensors with special characteristics. Examples of that are the ability to be humidity sensitive [10] or insensitive [2,11], the ability to sustain high temperatures [2,8,12], biological affinity [13], better mechanical properties [14], etc. Furthermore, as we have seen in the previous chapter, the possibility of doping POFs with special photosensitive materials such as fluorescein [1], trans-4-stilbenemethanol (TSB) [15], benzildimethylketal (BDK) [16], 9-vinylanthracene (9-VA) [17], diphenyl disulphide (DPDS) [18], etc., led to the inscription of FBG with high reflectivity [19,20] and with short inscription times [18,19,21]. POFBGs are nowadays not only fabricated at the 1,550 nm region but also at shorter wavelengths namely 960 nm [22], 850 nm [23], and 600 nm [24,25] where most of the polymer materials have lower attenuation. Multiplexing capabilities of these structures were shown to be possible through different techniques such as the inscription with different phase masks [26], the use of a single-phase mask and different POF elongations during inscription [27,28], the use of a single-phase mask with several diffraction orders [24] and the annealing with temperature [29], strain and temperature [30], and mediated by solution [31]. Intense research during the last five years revealed the possibility of fabricating not only uniform FBGs but also tilted [32], chirped [33], phase-shifted [34], and Fabry-Pérot (FP) cavities [35], opening the possibility of using these fiber optic devices for new emerging sensing applications. The possibility to inscribe long-period gratings (LPGs) in POFs also opens an easy route for low-cost POF sensors and it will also be included in this chapter.

3.2 MULTIPLEXED POFBG GRATING ARRAY

FBGs are ideal candidates for the detection of multiple sensing parameters. Therefore, the multiplexing capabilities together with the opportunity to create a quasi-distributed sensor are very promising. To achieve that, FBGs may be inscribed with different periods at different locations along the length of a fiber. This can be accomplished using the methods described in the previous chapter, namely: the interferometric method, the point-by-point (PbP) and line-by-line (LbL) methods and the phase mask method. Interferometric fabrication methods offer flexibility on the ability to produce FBGs in different wavelengths. However, the technique is susceptible to mechanical vibrations, submicron displacements in the position of the mirrors, air currents in the laser path and also require a laser source with good spatial and temporal coherence and with excellent output power stability. The PbP/LbL method offers also great flexibility on the fabrication of FBGs with different periods since the grating structure is built a point at a time. An example of the use of this technology in POFs may be found in Ref. [36] where an FBG array is reported at the end of a 6 m length GI-POF, at the 1,550 nm region, in which the grating separation was set to be higher than 8 cm each. The fabrication of dual FBG, separated by 3 m and operating at the 850 nm region, was also achieved in this fiber and the characterization to different parameters was performed as a proof of concept of a multiplexed POFBG sensor. This type of process can take up to 7 minutes for the inscription of a single Bragg grating [37]. In this way, the PbP method can be considered a time-consuming process. Additionally, errors in the grating spacing due to thermal effects and/or small variations in the fiber strain can occur during the inscription process, limiting the grating to short lengths. Furthermore, submicron translation and tight focusing limits make the task challenging to achieve first-order gratings.

On the other hand, the phase mask technique is one of the most effective methods to inscribe FBGs. Basically, it is only necessary the use of an optical element (i.e. phase mask), providing a robust and inherently stable method for reproducing FBGs. Furthermore, problems associated with mechanical vibrations or low temporal coherence of the laser beam are not problematic as opposed to interferometric methods. Based on this discussion, when flexibility in design is not needed, the phase mask method provides the best results. Therefore, it has been the preferred choice to inscribe FBGs in both silica and polymer fibers [38]. However, the lack of flexibility on the phase mask method led different authors to develop different techniques to permanently tune the Bragg wavelength. Some of those methods have already been developed for gratings written in silica fibers [39] and replicated in POFBGs [27], while others have been developed for the first time. A historical overview of the occurrence of these techniques may be seen in Figure 3.1.

A wide range of techniques has been used throughout the years to inscribe arrays of FBGs, and they will be briefly described in the following subsections.

3.2.1 INSCRIPTION WITH DIFFERENT PHASE MASKS

The simplest way to inscribe an array of FBGs using the phase mask technique is through the use of phase masks with different periods. Along the years, the inscription of Bragg gratings in POFs has been demonstrated at different wavelength regions, including 1,550 nm [1], 1,450 nm [41], 1,304 nm [42], 960 nm [22], 850 nm [23], and 600 nm [25]. Therefore, the sequential inscription of FBGs in a POF is rather simple and has been reported in Ref. [26] for the inscription of four FBGs in an undoped PMMA microstructured polymer optical fiber (mPOF) at the

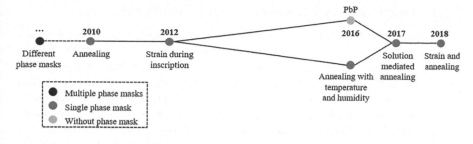

FIGURE 3.1 Historical diagram of the first report of a POFBG arrays using technologies, such as different phase masks; annealing with temperature [29]; strain during inscription [27]; annealing with temperature and humidity [40]; solution mediated annealing [31]; and annealing with strain [30].

infrared region. This work reported the use of a Krypton-Fluoride (KrF) UV laser operating at 1 Hz repetition rate (R), 0.5 mJ energy (E) and through the use of phase masks with periods equal to 1,023, 1,033, 1,061, and 1,072 nm. A schematic of the POFBG array may be seen in Figure 3.2a, while the reflected spectrum after the inscription in Figure 3.2b.

In order to increase the capacity of POF communication networks using wavelength division multiplexing (WDM) systems, Sáez-Rodríguez et al. implemented the fabrication of a coarse WDM system based on two overlapped POFBGs with different phase masks periods [43]. For that, the authors used an undoped PMMA mPOF and a 325 nm Helium-Cadmium (HeCd) laser. The first POFBG was written at the 850 nm region through a phase mask with a period of 570 nm, while the second grating was spatially overlapped over the former, using a phase mask with period of 1,066 nm, allowing to achieve a Bragg grating located at 1,550 nm region.

Despite the simplicity of swapping between phase masks, the method is costly and thus, other technologies have been proposed to suppress this problem.

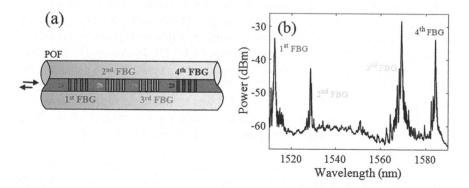

FIGURE 3.2 Inscription of a Bragg grating array in an undoped PMMA mPOF with $R = 1$ Hz and $E \sim 0.5$ mJ following reference [26]. Scheme of the implemented POFBG array (a). Reflection spectrum of the four POFBGs (b).

3.2.2 MULTIPLEXING POFBGS THROUGH ANNEALING PROCESS

When designing a POF sensor, it is important to know that fibers drawn under low temperature will present a high degree of birefringence, low ductility, high yield strength and tensile strength, and low thermal and strain stability [44–46]. This occurs due to the high degree of chain alignment along the fiber length. In fact, drawing a POF with low temperature will require a high drawing force leading to the formation of a high degree of chain alignment. However, if a thermal treatment is given to the fiber drawn under such conditions, it is possible to reverse the process and achieve characteristics that are found in fibers drawn under high temperatures/ low drawing forces. The thermal treatment, also designated by annealing, consists of heating the fiber for a determined period of time which depends on the degree of chain alignment present in the polymer. In most of the works reported to date, this can be up to 24 hours [8]. The annealing temperature shall be high but below the glass transition temperature (T_g) of the material present in the fiber. By using proper thermal annealing, the fibers can show low birefringence, high ductility, high thermal and strain stability, at a cost of fiber length decrease, diameter increase, and low yield strength and tensile strength [44,45]. Nevertheless, the use of high levels of humidity during thermal treatment improves the polymer molecules mobility, boosting the fiber shrinkage process [40,47]. Recently the use of methanol in water solutions at room temperature has also been used to achieve similar results to the annealing process with temperature and humidity [40].

Based on the influence that the annealing process has on the performance of the final POFBG device, the common methodologies used so far will be briefly reviewed.

3.2.2.1 Annealing with Temperature

The first work reporting the influence of temperature on POFBGs was performed by Carroll et al. [46]. The work demonstrated that heating a POFBG even with modest temperatures, can lead to a permanent blue-shift in the Bragg wavelength. This result is associated with a permanent shrinkage of the fiber and inherently, to a decrease of the period of the grating. Furthermore, authors were able to report a permanent blue wavelength shift of more than 30 nm by heating a PMMA-mPOF containing an FBG, up to 90°C. Later, the same group took advantage of this property to report an FBG WDM sensor in a PMMA multimode (MM)mPOF [29]. In this study, the authors wrote the first grating in a non-annealed fiber and then, performed an annealing process by exposing the POFBG to 80°C for 8 hours. The result was a permanent blue wavelength shift of ~18 nm. Another POFBG was then written with the same phase mask in another section of the annealed fiber, allowing the creation of two Bragg gratings, separated by 18 nm. The technique was later used by other authors to create dual parameter sensors [48].

After the work developed by Carroll et al., annealing POFs at high temperatures before or after the inscription of Bragg gratings become a common practice. By doing that, a POFBG sensitivity improvement in terms of linearity [45,49,50] and operational temperature [2,12], is easily achieved.

3.2.2.2 Annealing with Temperature and Humidity

Polymer materials such as PMMA, PC, polystyrene (PS), and others, present high water absorption capabilities. Until 2016, the annealing of fibers made of those materials was performed without controlling the humidity of the environment. At that time, two different groups showed the large shrinkage effect on hygroscopic POFs [47] and POFBGs [40] under controlled temperature and humidity conditions. Regarding specifically the work performed in Ref. [40], it was demonstrated that the annealing of POFBGs at a temperature of 80°C and 90% of humidity could induce a large blue wavelength shift of the resonance Bragg wavelength of more than 230 nm (see Figure 3.3a and b). Compared with the permanent wavelength shift reached by Carrol et al. [46], this wavelength shift was almost 8 times higher [40].

From Figure 3.3a and b, it can be seen that the resonance Bragg wavelength is blue-shifted from the 850 nm region to the low loss region of PMMA (~600 nm), reaching a permanent wavelength shift record of ~235 nm in about 24 hours [40], paving the way for a more intense use of the POFBG technology at the 600 nm region (where phase masks are more expensive and less effective). Furthermore, the capability to permanently shift a POFBG with such large range opened a new route for the development of coarse WDM systems for communications and sensors using just a single phase mask.

The effect of humidity on the annealing process described in Refs. [40,47] is explained by the plasticizer effect of water on the polymer material, which lowers its glass transition temperature [51]. Additionally, the polymer molecules mobility increases, leading the fiber to experience a large shrinkage similar to when the temperature is raised. Considering that the works described in Refs. [40,47] reported the annealing at high temperatures (~80°C–90°C), this effect is boosted [51].

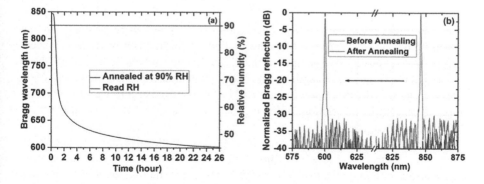

FIGURE 3.3 (a) Bragg wavelength shift during time, for a POFBG annealed under 80°C temperature and 90% humidity, during 24 hours. (b) Corresponding spectra before and after the annealing process. (Adapted from G. Woyessa, et al., "Temperature insensitive hysteresis free highly sensitive polymer optical fiber Bragg grating humidity sensor," *Opt. Express*, vol. 24, no. 2, pp. 1206–1213, 2016, with permission from the Optical Society of America.)

3.2.2.3   Solution Mediated Annealing

Early studies developed by Smith and Schmitz [52] demonstrated that when PMMA is at equilibrium with water, the T_g of the polymer is lowered by approximately 20°C when compared to the one found in dry conditions. Other polymer-solvent combinations may also apply, being the T_g of the polymer dependent on the polymer and solvent/solution interaction. One example is the use of methanol in an equilibrium polymer-solution system, which has shown to reduce the T_g of PMMA down to the ambient temperature [53]. Based on this effect, Fasano et al. created a new way to relax the stresses frozen in a PMMA POFBG by immersing the fiber in different concentrations of methanol/water solutions at room temperature [31]. The results showed different degrees of fiber shrinkage depending on the solution concentration. This behavior was associated with the different degrees of relaxation of the polymer chains frozen in stress after the heat-drawing process. Consequently, a permanent blue wavelength shift was observed, obtaining similar results to the ones achieved for conventional annealing methods reported in Ref. [40] for POFBG annealed simultaneously with temperature and humidity. Since this methodology does not require the use of a climatic chamber, it can be considered a cost-effective method to permanently tune the Bragg wavelength and thus to create WDM systems in POF using a single phase mask.

3.2.3   POFBG-WDM Fabrication with Stress during Inscription

One of the oldest methods to write Bragg gratings with different wavelengths using a single phase mask was reported in 1994, by controlling the fiber stress during the inscription process [39]. In fact, the resonance Bragg wavelength can be precisely tuned by controlling the stress imposed on the fiber, through the following expression:

$$\Delta\lambda_{\text{Bragg}} = \lambda_{\text{Bragg}}(1 - p_e)\varepsilon \qquad (3.1)$$

where p_e defines the photoelastic coefficient of the fiber material, λ_{Bragg} is the resonance Bragg wavelength, and e is the applied strain. In this way, if one considers to apply stress during the inscription, the created Bragg wavelength will still have the resonance Bragg wavelength given by the phase mask period. However, when stress is released, the Bragg wavelength will be blue-shifted accordingly to the strain imposed during the inscription, following Equation 3.1. A schematic of this procedure may be seen in Figure 3.4.

While the described technique could be applied for both silica and polymer fibers, it is interesting to note that polymer fibers can recover strains up to 4.7% depending on the strain rate [54]. This value is much higher than the one found for uncoated silica fibers, that is, approximately equal to 0.6% [55]. In this aspect, POFBGs offer a clear advantage. Related to this topic, Yuan et al. decided to inscribe POFBG kept under strain, allowing to report a dual-POFBG temperature compensated strain sensor [27]. Their results revealed the possibility to permanently linear tune the Bragg wavelength up to 8 nm by straining the fiber during the inscription with values below 1%. Above that point, the tuning curve saturated showing a tuning record

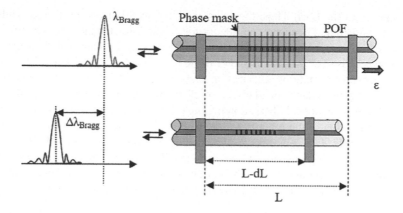

FIGURE 3.4 Schematic of the fabrication of an FBG through the phase mask method and applying strain during the inscription (top-right image). Corresponding shrinkage of the fiber after strain release (bottom-right image). The correspondent Bragg wavelength is blue-shifted when the strain is released as seen on the left part of the image.

of 12 nm for the inscription under a strain value of 2.25%, which was larger than the 2.5 nm Bragg wavelength shift reported for silica FBGs [39]. The same principle was later reported by other authors to report multiple POFBGs in a SI-POF [28] and also a WDM system with overlapped FBGs [43].

Compared with the annealing technique, FBGs fabricated in fibers kept under strain are much more controllable and the process can be applied for POFs regardless of their drawing conditions and whether they have been annealed or not. Yet, both techniques can only tune the Bragg wavelength to shorter wavelengths. Nevertheless, according to the results presented by Pospori et al., it is possible to permanently tune the Bragg wavelength to longer wavelengths by annealing POFBGs under different stretching conditions [30]. This behavior can be explained because the polymer chains cannot move freely due to the stretching conditions, and thus, only small molecular conformations may occur during this stretching-annealing process. As the fiber is cooled down to the ambient conditions, the Bragg wavelength shifted by the stretching conditions becomes permanent. Using this technique, Pospori and co-workers were able to permanently red-shift the resonance Bragg wavelength of a PMMA-mPOF up to 10.4 nm by annealing the fiber in hot water at 50°C and with a constant stress of ~13 MPa. Using this technique, the authors were able to create an FBG array by controlling the strain conditions during the annealing process.

3.3 BRAGG GRATINGS IN BIREFRINGENT POFs

In an ideal circular-core fiber, two orthogonally polarized modes propagate with the same phase velocity. However, practical fibers don't have a perfectly circular symmetric core and because of that, each mode propagates with slightly different phase and group velocities. Additionally, external factors such as bend, twist, anisotropic stress, temperature, and humidity (in the case of hygroscopic materials) also introduce birefringence in the fiber. Thus, the transfer of energy from one polarized mode to the other can occur. Therefore, the magnitude of the birefringence can change over time depending on the environment conditions, creating problems in field applications. Sensors based on FBGs are one of those cases, as the dependence of the Bragg wavelength with the state of polarization may induce errors in the parameter being measured. Despite its detrimental effect, birefringence can be intentionally added in a controlled way to an optical fiber, allowing the creation of fibers that can propagate the state of polarization of the input light for long distances, also known as polarization maintaining (PM) fibers. One specific fiber found in this category is the HiBi fiber, in which the propagation constants of two orthogonally polarized modes are quite different and thus the coupling between each other is greatly reduced. The fibers ability to maintain its polarization state is described by the beat length and it is defined as:

$$L_B = \frac{2\pi}{\Delta\beta} = \frac{\lambda}{\Delta n_{\text{eff}}} \qquad (3.2)$$

where $\Delta\beta$ is the difference between the propagation constants for the two orthogonally polarized modes, associated with the fast and slow optical axis of the fiber and Δn_{eff} is the difference between the effective indices for each polarized mode at the wavelength λ. L_B defines the distance along the fiber required to bring the two polarization modes back in phase to the original polarization state.

3.3.1 BIREFRINGENCE INDUCED BY LASER LIGHT

Birefringence in POFs was first reported in 2001 when Zhang and co-workers fabricated a long period birefringent grating in a multimode azobenzene doped PMMA POF using polarized light from a 532 nm Nd:YVO4 laser [56,57]. The phenomenon was attributed to the photo-induced alignment of the azobenzene units (Figure 3.5) allowing to achieve a maximum induced index change of 3×10^{-5} [56,58].

Subsequent works performed in a single-mode POF (SM-POF) with core composed of azobenzene copolymer instead of azobenzene doped polymer and using the amplitude mask technique demonstrated an induced 120 μm birefringent LPG [59,60]. The refractive index change in the exposed area was about 1×10^{-3} [61].

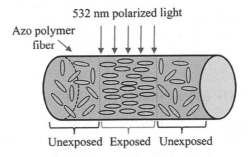

FIGURE 3.5 Photoinduced alignment of the azo units in the POF when it is exposed to polarized light at 532 nm visible radiation [59].

Photo-induced birefringence in POFBGs resulting from side exposure of SI-POF to UV radiation from a continuous HeCd UV laser operating at 325 nm, has been the subject of study in 2014 [62]. Results revealed that the amount of birefringence (i.e., 7×10^{-6} [62]) was one order of magnitude higher than the intrinsic fiber birefringence, confirming that the side inscription process creates a non-negligible photo-induced birefringence, as it occurs with silica fibers [63]. Further studies performed by the same authors revealed that the photoinduced birefringence in FBGs written in mPOFs was higher (3×10^{-5}). This difference was explained due to the microstructured nature of the fiber which creates more photo-induced defects during the UV irradiation [64].

The amount of induced birefringence during the FBG inscription process can be advantageous for the development of temperature-insensitive transverse force sensing devices. Inspired by the work developed in silica FBGs [63], Hu and co-workers investigated the response of a SI-POFBG to lateral force, for various radial fiber directions using the polarization-dependent loss (PDL) response. By doing that, the authors were able to report a maximum sensitivity of 2.57 dB/N with a temperature insensitivity (<3% error for a 10°C variation) [64].

3.3.2   FIBER BRAGG GRATING FABRICATION IN HiBi mPOF

During the fiber production process, birefringence can be deliberately introduced through the creation of an asymmetric strain on the fiber core region, allowing to obtain an asymmetry on the index profile. This can be done through the introduction of highly doped regions around the core. Other possible methods may be done by breaking the circular symmetry of the fiber using elliptical fiber core shapes. Nevertheless, the exceptional design flexibility on tailoring the light transmission in an optical fiber using microstructured fibers opens a range of possible configurations by just controlling the number, the size, the position, and the shape of the air holes surrounding the core region. Microstructured polymer optical fibers bring new opportunities over their silica counterparts due to the low processing temperatures and the easiness on the fabrication of the fibers preforms which can be obtained through the use of different technologies, such as preform casting [65], drilling [66], extrusion [67], or more recently, 3D printing [68]. Because of these advantages, it is not surprising to find the first HiBi mPOFs in the early days of the mPOF technology in 2004 [69]. A number of subsequent works were then followed using different hole arrangements [66,70,71].

Fiber Bragg gratings written in HiBi fibers have interesting properties involving the possibility to reflect different wavelengths for each polarization mode. Hence, some sensing applications have been proposed, such as the capability to measure simultaneously longitudinal and transverse strain and temperature [72]. Based on those properties and taking into account the mechanical properties given by polymer materials, Oliveira et al. reported the fabrication of high-quality Bragg gratings in HiBi-mPOFs [6]. For that, the authors used two undoped PMMA HiBi-mPOFs shown in Figure 3.6a and b.

For the Bragg grating inscription, the authors used the 248 nm phase mask recording system described in Ref. [73]. Two Bragg gratings were written in each fiber with

FIGURE 3.6 HiBi mPOFs made of undoped PMMA, used for the production of FBGs in Ref. [6].

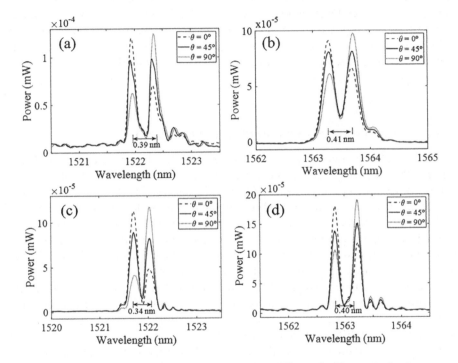

FIGURE 3.7 Linear reflection spectra obtained at different incident polarization states, obtained for two different wavelength regions (~1,520 and ~1,560 nm). Images shown in ((a), (c)) and ((b), (d)), correspond to the inscription with 1,033 and 1,062 nm phase mask periods, respectively. The images shown in ((a), (b)) and ((c), (d)) are referred to HiBi mPOF shown in Figure 3.6(a) and (b), respectively. (Reprinted from R. Oliveira, et al., "Bragg gratings inscription in highly birefringent microstructured POFs," *IEEE Photonics Technol. Lett.*, vol. 28 (2016), no. 6, pp. 621–624, 2016, with permission from IEEE© (2016).)

phase mask periods of 1,033 and 1,062 nm, creating two Bragg gratings at different wavelengths. Their growth was monitored in reflection using an interrogator and it took only a few laser pulses for the inscription, allowing to avoid saturation. After the inscription, two clearly separated Bragg peaks ($\Delta\lambda_{\text{fast,slow}} \approx 0.3 - 0.4$ nm), appeared at the wavelengths that satisfied the Bragg condition (see Figure 3.7), such that:

$$\lambda_{\text{Bragg,slow}} = 2n_{\text{eff,slow}}\Lambda$$

$$\lambda_{\text{Bragg,fast}} = 2n_{\text{eff,fast}}\Lambda \tag{3.3}$$

where Λ denotes the period of the grating. The results revealed that the phase bire-fringence, calculated through:

$$\Delta\lambda_{\text{fast,slow}} = \lambda_{\text{Bragg,fast}} - \lambda_{\text{Bragg,slow}}$$

$$= 2n_{\text{eff,fast}}\Lambda - 2n_{\text{eff,slow}}\Lambda$$

$$= 2B\Lambda \tag{3.4}$$

$$\Leftrightarrow B = \frac{\Delta\lambda_{\text{fast,slow}}}{2\Lambda}$$

was similar to the ones estimated through numerical simulations and reached values of the order of 10^{-4}. The interrogation of the POFBGs for different input light polar-ization states revealed clear evidence of a true PM behavior as shown in Figure 3.7a–d for the mPOFs shown in Figure 3.6a and b, respectively.

3.4 TILTED BRAGG GRATINGS IN POFs

Tilted fiber Bragg gratings (TFBGs) are short-period gratings (grating period close to 500 nm for the use in the C+L bands) that have all the advantages presented for the well-developed FBG technology together with the ability to excite cladding modes resonantly. In TFBGs, the fiber refractive index is modulated by a tilt angle (θ_{TFBG}) with respect to the fiber transversal axis (inset shown in Figure 3.8). In these gratings, two couplings may occur the self-backward coupling of the core mode and numerous backward couplings between the core mode and different cladding modes [74]. Hence, TFBGs can show all the sensing opportunities of the conventional

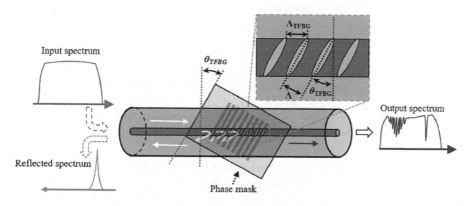

FIGURE 3.8 Schematic representation of a tilted fiber Bragg grating fabrication and principle of operation.

FBGs and also provide new ones due to the sensitivity of the cladding modes to the external environment. Thus, potential applications, such as bend sensors [75] or refractometers [76], may be explored. Furthermore, the presence of the Bragg wavelength provides the additional benefit of having a convenient temperature reference. When TFBGs are seen in transmission, as represented in Figure 3.8, they show one resonance at longer wavelengths associated with the core mode and a comb-like transmitted amplitude spectra at shorter wavelengths composed of several tens of narrow-band resonances, associated with the cladding modes.

The resonance Bragg wavelength λ_{TFBG} and the ith cladding mode resonances $\left(\lambda_{\text{clad}}^{i}\right)$ can be determined by the phase-matching condition, through the following expressions:

$$\lambda_{\text{TFBG}} = \frac{2n_{\text{eff,core}}\Lambda}{\cos\left(\theta_{\text{TFBG}}\right)} \tag{3.5}$$

$$\lambda_{\text{TFBG}}^{i} = \frac{\left(n_{\text{eff,core}}^{i} - n_{\text{eff,clad}}^{i}\right)\Lambda}{\cos\left(\theta_{\text{TFBG}}\right)} \tag{3.6}$$

where $n_{\text{eff,core}}$ is the effective index of the core mode at λ_{Bragg} while $n_{\text{eff,core}}^{i}$ and $n_{\text{eff,clad}}^{i}$ are the core mode and the ith cladding mode at $\lambda_{\text{clad}}^{i}$, respectively [76]. θ_{TFBG} is the tilt angle of the grating planes related to the fiber transversal axis, Λ is the nominal grating period, described as $\Lambda = \Lambda_{\text{TFBG}}\cos(\theta_{\text{TFBG}})$, being Λ_{TFBG} the grating period along the fiber axis.

TFBGs in silica fibers is a well-developed area with many different applications [77]. Nevertheless, the possibility of using this type of gratings in POFs brings additional opportunities, for instance, in curvature and biochemical sensors, due to the improved flexibility and biological compatibility of POFs, respectively. While the first POFBG had been demonstrated in 1999, the fabrication of TFBGs in POFs was only shown fifteen years later [32]. The reason for that was essentially due to the necessity of higher power for the photo inscription of TFBGs when compared with uniform FBGs. This occurs due to the fact that the coupling efficiency of the grating decreases due to its angled fringes, as reported in Ref. [74]. To suppress the problem of low laser power on the TFBG inscription in POF, Hu and his co-workers [32] used the methodology described in their prior work [62], that consists on the phase mask method and a slightly etched photosensitive SI-POF (PMMA core doped with diphenyl sulfide (DPS) (5% mole) and TSB (1 wt%)). The TFBG was written using a 325 nm, 30 mW HeCd laser with 1.2 mm spot diameter and the phase mask was slightly angled related to the transversal optical fiber axis: $\theta_{\text{TFBG}} = 1.5°$, $3.0°$, and $4.5°$. By doing that, they were able to get the spectra shown in Figure 3.9 [32].

The results reported in Ref. [32] and shown here in Figure 3.9, led the authors to have the same observations taken for silica FBGs, since the increase of the tilt angle led to increase the spectral content, due to the coupling to higher-order cladding mode resonances; as well as an expected red-shift of the Bragg wavelength due to

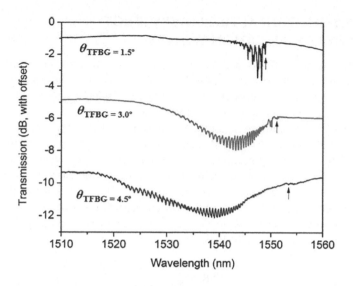

FIGURE 3.9 Transmission spectra obtained from a POF-TFBGs written with 1.5°, 3.0°, and 4.5°, in a slightly etched, photosensitive SI-POF. (Reproduced from X. Hu, et al., "Tilted Bragg gratings in step-index polymer optical fiber," *Opt. Lett.*, vol. 39, no. 24, pp. 6835–6838, 2014, with permission from the Optical Society of America.)

the phase-matching condition of the fundamental mode (Equation 3.5). Also, the peak-to-peak amplitude of the cladding modes decreased with the increase of the tilt angle, which was in accordance with Ref. [74].

Hu et al. also characterized one of the grating structures ($\theta_{TFBG} = 3.0°$) to the surrounding refractive index (1.42–1.49), obtaining a sensitivity of ~13 nm/RIU, which was similar to that of silica TFBGs.

Surface plasmon resonance (SPR) is another technology widely used in refractometry due to the high performance in terms of sensitivity [78]. It is an accurate technique capable to determine density changes at the interface between a metal and a dielectric medium (or analyte). The improved biocompatibility of POFs compared to silica fibers for future in-vivo applications opens new possibilities for the development of POF-SPR based sensors. In literature, it is possible to find results reporting the use of SPR in tapered and unclad MM-POFs at the visible region [79]. However, working at the infrared region allows higher penetration depth of the evanescent field to the outside medium. TFBGs have the ability to break the optical fiber circular symmetry, allowing a straightforward coupling of the light to radial modes required to excite the SPR modes. Thus, Hu et al. [80] based on the knowledge gathered on the inscription of TFBGs in POFs [32] and taking into account the biocompatibility advantages of these fibers as well as the benefit in working at the infrared region, deposited a 50 nm gold layer on a 6° tilted POFBG. By doing that, the authors were able to characterize the fiber structure to the external refractive index, allowing to get a refractometric sensitivity of ~500 nm/RIU, which is comparable with other plasmonic optical fiber configurations [78].

3.5 POLYMER OPTICAL FIBER FABRY-PÉROT CAVITIES FABRICATION WITH FBGs

The easiness to write FBGs with specific reflectivity makes them ideal for their implementation as reflective mirrors in Fabry-Pérot cavities. These structures were first reported by Huber in 1991 [81] and consisted of two FBGs spaced by a small distance, forming a resonant cavity as shown in Figure 3.10.

FP cavities work similarly to bulk FP interferometers except that the gratings are narrowband and are distributed reflectors. Due to the presence of the FP cavity between the two Bragg gratings, the original reflection spectrum of the grating is modulated by the interference fringes, whose depth increases while the width decreases with exposure time. One of the parameters that characterize this type of filters is the free spectral range (FSR), which defines the spacing between maxima of the spectral response of the cavity. For an FP cavity formed by two mirrors, separated by a distance d, the FSR may be expressed as:

$$FSR = \frac{\lambda^2}{2dn_{eff}(\lambda)} \tag{3.7}$$

where λ defines the free-space wavelength and n_{eff} the effective refractive index. For an equivalent FP cavity formed by two Bragg gratings, the FSR cannot be calculated from a single distance d. In fact, d becomes a function dependent on the wavelength. In these conditions, the FSR is maximum on the reflection peak and can be calculated in good approximation by Equation 3.7 and considering d as the distance between the inner ends of the cavity.

FP cavities based on FBGs have a set of very narrow filters with applications in optical communications and sensors due to its simplicity, small size and increased measured resolution. As it is known, POFs have a much lower Young's modulus than silica fibers and thus, the possibility to write such structures in POFs is of special interest for optoacoustic endoscopy [82,83]. Nevertheless, POFBGs with good stability are normally obtained from gratings of type II. However, these types of gratings achieve bandwidths of the order of nanometer limiting their use in high-resolution sensing applications [35].

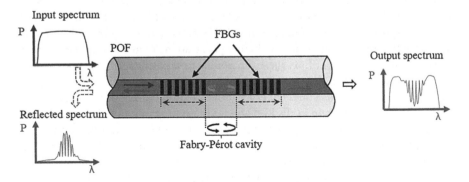

FIGURE 3.10 Schematic representation of a Fabry-Pérot cavity formed by two closely separated FBGs.

The first description of an FP cavity fabricated through the use of polymer fiber Bragg grating was reported in 2005 by Dobb and co-workers [84,85]. The gratings were written at the 1,550 nm region in a PMMA SI-POF through the phase mask technique, using an HeCd laser operating at 325 nm. The length of the gratings was 1 cm and the separation between them was 3 cm, allowing to achieve a spectrum profile with an approximate sinusoidal shape. However, the spectral profile ressembled a low finesse cavity, which according to the authors, was due to the high attenuation of the POF at the infrared region.

In 2016, Pospori et al., interested in the potential of POFs for stress sensing applications, reported a numerical model to study the performance of a POF-based FP sensor formed by two uniform Bragg gratings with finite dimensions [86]. Results have shown that the stress sensitivity was higher for gratings with higher reflection coefficients and lower cavity lengths. Furthermore, the attenuation of the fiber revealed to deteriorate the performance of the proposed device. However, due to the different material properties of silica and polymer materials, the stress sensitivity for the case of FP cavities based on POFBGs could still find regimes where the sensitivity is better than that of the silica optical fiber.

Recently, Statkiewicz-Barabach et al. [35,87] reported an interesting work regarding the possibility to fabricate FP structures with very sharp fringes in a PMMA SI-POF with core made of PMMA/PS copolymer. The creation of the fiber structure was done at the 1,300 nm wavelength and it was made in a single step, using an HeCd laser through the phase mask technique and placing an "obstacle" (e.g., metal wire), before the phase mask and at the middle of the irradiated area, as represented in Figure 3.11.

By exposing a 1.7 mm UV radiation through a ~100 μm diameter metal wire, placed at the center of the phase mask, the authors were able to report an FP cavity with FSR of about 1.7 nm, corresponding to a Fabry-Pérot cavity length of 340 μm [35]. The fabricated structure had very sharp fringes with linewidths ranging between 50 and 100 pm, allowing the authors to get a sensing resolution 15 times higher than a standard POFBG.

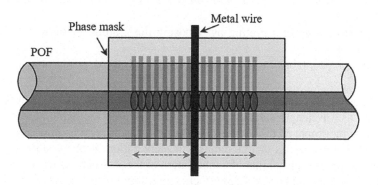

FIGURE 3.11 Methodology used in Ref. [35] for the fabrication of an FP cavity. The fabrication is based on the inscription of a single Bragg grating with a blocking aperture positioned in the middle of the phase mask.

The fabrication of FP cavities based on FBGs has also shown to be possible in multimode perfluorinated GI-POF (the most transparent POF) [88]. To do that, a direct-write PbP laser inscription methodology has been implemented, through a femtosecond laser operating at 517 nm, emitting 220 fs pulses at 2 kHz repetition rate and with pulse energies of ~80 nJ. To reduce the number of fiber modes coupling to the grating, due to the MM nature of the fiber, the authors limited the FBGs spatial extent to the central part of the core where the gradient index profile peaks. In this way, they were able to excite the strongest lower order modes, allowing to get single peak POFBG excitation [37]. The FP cavity was thus demonstrated by inscribing two 4th order POFBGs with lengths of 660 μm, separated by a distance of 3 mm and with plane widths of 15 μm in the central region of the fiber [88]. The result was an FP cavity with a FSR of 240 pm and a finesse of 4, indicating POFBG reflectivity of about 65%.

3.6 PHASE SHIFTED POLYMER OPTICAL FIBER BRAGG GRATINGS

Phase shifted Bragg gratings may be considered as a particular example of FP cavities based on FBGs, where the distance between the inner ends of the cavity are lower than the modulation period. A phase jump opens up a bandgap within the reflection bandwidth, creating a narrow transmission band, where the bandwidth can reach values of the order of a few tens of picometers. The procedure is well known and has been applied in distributed feedback lasers to allow stable SM operation [89]; applications in the microwave photonics field, such as a tunable bandstop-to-bandpass microwave photonic filter [90]; and sensors with increased resolution compared with standard FBGs [91].

A phase-shifted grating is, in its simplest form, an FBG that contains a phase shift across the fiber grating, whose location and magnitude may be adjusted to design a specific transmission spectrum. The introduction of a phase shift may be performed through different methods, such as [38]:

a. Using a phase mask, in which the phase shift regions have been written into the mask design;
b. Post-processing of a grating by exposing the grating region to UV laser radiation;
c. Through post-fabrication processing using a localized heat treatment.

These types of procedures lead to the creation of two gratings that are out of phase, which acts as wavelength-selective FP resonator, allowing the light at the resonance to penetrate the stopband of the original grating.

The special properties of POFs make the fabrication of phase-shifted gratings very appealing. Because of that, the fabrication of these structures has been demonstrated by different authors, different technologies and different spectral windows, along the years [34,84,85,92,93]. The first work reporting a phase shifted POFBG was presented in 2005 by Webb et al. [84]. The authors first wrote a 1 cm standard uniform grating at the 1,550 nm region in a PMMA SI-POF, using the phase mask

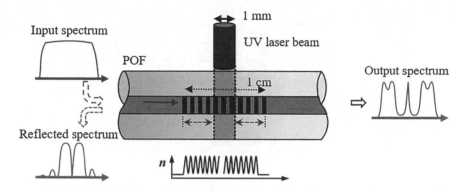

FIGURE 3.12 Methodology used in Refs. [84,85] to introduce a phase shift in a POFBG by increasing the refractive index at the center of the grating through UV exposure.

technique through an HeCd laser operating at 325 nm region. Then, the center of the grating was exposed directly to the output of the laser beam (diameter of ~1 mm) allowing the increase of the refractive index in the exposed region. A schematic procedure of the technique may be seen in Figure 3.12.

The grating at the region of the spot of the UV beam was partially erased, allowing the creation of a π-shifted grating. Despite the great achievement on the report of this kind of structures in POF, the output reflection spectrum revealed lack of sharpness in the central notch, which according to the authors was due to the high attenuation of the fiber material at the 1,550 nm region [84].

Seven years later, Zhou et al. took advantage of the low loss of TOPAS® material at the terahertz (THz) region to report a narrowband notch filter using a $\pi/2$ phase-shifted FBG in a 600 μm unclad TOPAS® based POF [92]. To accomplish that, the authors used the technology described in their earlier work [94], in which a 248 nm KrF UV excimer laser, with pulse energy of 200 mJ and repetition rate of 5 Hz, was used to record a Bragg grating structure with the PbP technique. Two different phase shift Bragg gratings were formed by ablation, where each grating plane was formed by a triangular notch, in which the depth and width were dependent on the exposure time. The gratings length was 98 mm and had a total of 192 notches. The width of each grating plane was 200 and 250 μm while the depth was 240 and 278 μm [92]. The etched grooves were imprinted with periods of 509 μm with the exception of the center of the FBG, where a $\pi/2$ phase shift was inserted. The result was an FBG centered at ~260 GHz presenting a narrow transmission notch at the FBG peak.

Recently, it was reported the fabrication of a high-quality phase-shifted grating at the 850 nm region in a BDK-doped PMMA mPOF [34]. To do so, the authors implemented the methodology described in Ref. [73] (248 nm KrF UV laser and the phase mask technique) and the phase shift was created with the same approach implemented in Ref. [35], where a blocking aperture (metal wire of 40 μm in diameter) was used at the center of the 1 cm length phase mask, to block the UV radiation at the central region. The structure was fabricated with a single UV laser pulse thanks to the photosensitivity of the fiber at short UV wavelengths. The grating was observed in transmission, presenting two dips separated by 61 pm and with transmission losses

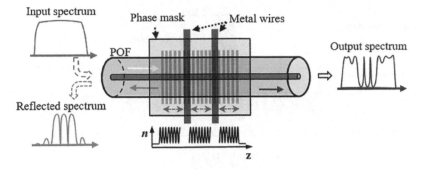

FIGURE 3.13 Fabrication methodology used in Ref. [34] to produce a POFBG with two-phase shifts through the use of two metal wires equally spaced along the length of the grating planes of the phase mask, during the inscription process.

of 16.3 and 13.2 dB for the first and second dips, respectively. The transmission notch depth was 10 dB (~15 pm at 3 dB), with an estimated phase shift of ~4π/3 [34]. Using the same approach but with two narrow wires equally spaced along the 10 mm length phase mask, the authors were able to report a POFBG with two-phase shifts, as represented in Figure 3.13. The fabricated structure presented two well-defined notches with depths of about 9.9 dB and separated by ~58 pm.

3.7 MOIRÉ POLYMER OPTICAL FIBER BRAGG GRATINGS

When two interference fringes with slightly different periods Λ_1 and Λ_2 are superimposed in fiber, the resulting Bragg grating will present a modulation of the refractive index with a Moiré effect. This effect results from the beat between the two similar spatial frequencies, originating a quick refractive index amplitude response envelop with period Λ_g, given by:

$$\Lambda_g = \frac{2\Lambda_1\Lambda_2}{\Lambda_1 + \Lambda_2} \tag{3.8}$$

and one slowly response envelop with a period Λ_e, expressed by:

$$\Lambda_e = \frac{2\Lambda_1\Lambda_2}{\Lambda_1 - \Lambda_2} \tag{3.9}$$

In this way, the resultant index modulation change along the longitudinal axis of the fiber ($\delta n_{\text{eff}}(z)$), can be described through [95]:

$$\delta n_{\text{eff}}(z) = \overline{\delta n_{\text{eff}}}\left[2 + 2\cos\left(\frac{2\pi z}{\Lambda_e}\right)\cos\left(\frac{2\pi z}{\Lambda_g}\right)\right] \tag{3.10}$$

In Figure 3.14, it can be seen a Bragg grating with the Moiré effect.

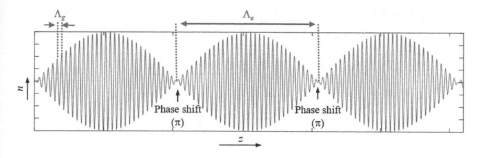

FIGURE 3.14   Example of a Bragg grating with Moiré effect.

The minima of the envelope correspond to π phase jumps. Spectrally this phase jump has a similar effect to the phase-shifted Bragg grating, which results on the appearing of a narrowband transmission filter at the rejection band of the grating. Phase shifts other than π may be achieved by the overlap of gratings with different amplitudes. Nevertheless, multiple phase shifts can be obtained either by using longer gratings or by changing Λ_1 and Λ_2 while the grating length is maintained [96].

The fabrication of this type of structures in POFs has been reported recently at the 1,550 nm region for a BDK-doped PMMA mPOF [93]. The authors used the phase mask technique through a 325 nm HeCd laser (2.5 mm beam diameter) and a single-phase mask with pitch equal to 1,066 nm. The methodology was based on previous works reported for the inscription of POFBGs closely separated [27] and spatially overlapped [43]. For that, the authors wrote the first grating with the POF under 1.00% strain, while the second grating was written in the same fiber location by slightly increasing the strain to 1.04%. After the inscription, the authors were able to report a phase shifted POFBG based on the Moiré effect. The structure was seen in reflection showing an 11 dB stopband with two sidebands separated at the central region by a notch with 5 dB insertion loss and bandwidth close to 200 pm. Later, the same group reported the fabrication of the same type of structure at the 850 nm wavelength region, where the polymer material offered lower losses. For that a 248 nm KrF UV laser [96] was used instead of a 325 nm HeCd laser, allowing them to also benefit from the lower inscription time [96].

3.8   CHIRPED FIBER BRAGG GRATINGS (CFBGs) IN POFs

Chirped fiber Bragg gratings (CFBG) have a monotonous varying period of the refractive index modulation along the grating length. The most important configuration is the linearly chirped FBG, where the refractive index period changes linearly along the grating length. In this type of grating, the ability to reflect different wavelengths is possible, resulting in a bandwidth much broader than the one found for uniform FBGs, ranging from few nanometers to tens of nanometers. Furthermore, since different wavelengths are reflected in different positions along the grating length, it results in a group delay dependent on wavelength. This property has many advantages in optical communications, namely by introducing a differential group delay dependent on wavelength [97].

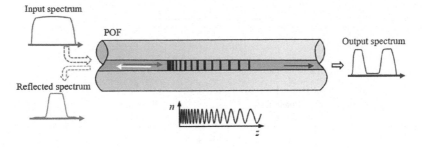

FIGURE 3.15   Operating principle of a chirped fiber Bragg grating, where the refractive index period is linearly increased.

Generally, the period of a CFBG can be expressed by a N degree polynomial function given by:

$$\Lambda(z) = \Lambda_0 + \Lambda_1 z + \cdots + \Lambda_N z^N \tag{3.11}$$

where Λ_0 and Λ_N is the period at the input and output of the grating, respectively. The operating principle of a CFBG may be seen in Figure 3.15, where a broadband source is injected onto an FBG with refractive index varying linearly along the grating length, allowing then to observe the reflection and transmission spectra with large bandwidth.

The interesting characteristics of CFBGs compared with uniform FBGs make them appealing for sensor applications [98]. In fact, the broadband spectrum of a CFBG can respond to external parameters such as strain and temperature as it does FBGs but with additional features, like the detection of localized events like temperature hot spots [99] or strain discontinuities [100], in addition to the ability to detect the values of those parameters.

The fabrication of CFBGs in silica fibers is well documented in the literature and relies on the use of different technologies, such as:

* Use of phase masks with chirp periods [101];
* Moving the fiber or phase mask related to each other during the inscription [102];
* Exposure to the interference pattern formed by two beams with dissimilar phase fronts [103];
* Bending the fiber in a uniform fringe pattern [104];
* Applying a gradient temperature along the length of the fiber during the inscription [105];
* Applying strain gradient along a fiber taper to produce the chirp either during or after writing a uniform grating [106];
* Use of phase masks that can be thermally tuned during the inscription process [107].
* Use of direct writing methods such as the PbP and LbL.

Some of these methodologies have already been explored for the production of CFBGs in POFs and thus, they will be briefly reviewed in the following subsections.

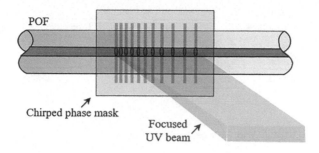

POF

Chirped phase mask

Focused
UV beam

FIGURE 3.16   Schematic representation of the inscription of a CFBG through the use of a chirped phase mask.

3.8.1   Phase Mask with Specific Chirp Period

The first proposal of a CFBG in a POF was made in 2003 [108], where Liu reported the first simulation results on the dynamic dispersion compensation using linearly chirped POFBGs under different strain conditions. Later, in 2005, the same author and co-workers [109] reported theoretical results on the tunable control of dispersion with CFBGs in POFs, with fixed center wavelength. Their results relied on the assumption that the tensile strain could adjust the dispersion while the temperature rise had an effect on offsetting the center wavelength induced by the applied strain. Simulation results revealed a very large dispersion tuning range (from 2,400 to 110 ps/nm) with potential applications for tunable dispersion in wavelength division multiplexing systems [109]. The POFBG technology has nowadays matured to a degree capable of producing these types of gratings, mainly due to the improvement on the photosensitivity of the POFs as well as the use of suitable inscription sources. In this way, the first CFBG based on POF was reported in an undoped PMMA SI-POF [33]. The authors used the phase mask technique through a 248 nm KrF pulsed UV laser and using a 25 mm long chirped phase mask with period of 1,068 nm and linear chirp of 1.2 nm/cm [73]. The laser beam was expanded to allow total exposure of the grating planes for each laser pulse. A schematic example of the technique may be seen in Figure 3.16.

After a few shots, the authors were able to report a CFBG based on POF at ~1,580 nm with a bandwidth of 3.9 nm, peak-to-noise level of 16.9 dB and with 3.8 dB ripple within the reflected bandwidth [33].

3.8.2   Tapering Method

While the use of a phase mask with a specific chirp period seems to be the most easiness and repeatable way to produce a CFBG, it is however quite expensive when different resonance Bragg wavelength or chirp periods are required. According to Cruz et al. [110], the fiber tapering technique [99] offers versatility and simplicity compared to other technologies and because of that, many works have been reported on this subject [98]. The technique relies on the inscription of a grating through a uniform phase mask, using a tapered fiber subjected to strain during or after the inscription, allowing to accurately control the chirp parameter. A schematic of the

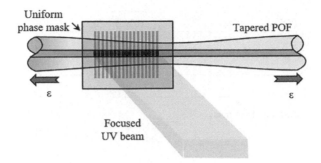

FIGURE 3.17 Schematic of the CFBG inscription process, using a tapered POF kept under strain during the irradiation through a uniform period phase mask.

inscription method through a tapered fiber kept under strain during the irradiation process may be seen in Figure 3.17.

The theory behind the fabrication of CFBGs in tapered fibers using strain may be explained by considering a uniform Bragg grating in an optical fiber of radius a, subjected to a stress F. The result will be a change on both the grating pitch and effective refractive index, leading to a change on the Bragg wavelength λ_{Bragg} according to:

$$\frac{\Delta\lambda_{Bragg}}{\lambda_{Bragg}} = \frac{F}{\pi E a^2} - p_e \frac{F}{\pi E a^2} \tag{3.12}$$

where E is Young's modulus of the fiber and p_e is the photoelastic coefficient of the fiber material. The first term of the equation denotes the physical length change and the second term denotes the refractive index change due to the strain optic effect. According to Equation 3.12, a non-uniform diameter fiber Bragg grating will reflect different wavelengths. Thus, it is possible to produce gratings with any arbitrary chirp by proper design of the fiber diameter along the grating. The chirp function $f(z/L)$ and the fiber profile $r(z)$ are related by [110]:

$$r(z) = \frac{r(0)r(L)}{\sqrt{\left(r(0)^2 - r(L)^2\right)f\left(\frac{z}{L}\right) + r(L)^2}} \tag{3.13}$$

where L defines the length of the grating. Thus, if a linearly chirped grating is considered, $f(z/L) = z/L$, the Bragg wavelength will be defined according to Ref. [110], as:

$$\lambda_B\left(\frac{z}{L}\right) = \lambda_{Bragg}(0) + \Delta\lambda_{Bragg}\frac{z}{L} \tag{3.14}$$

where $\Delta\lambda_{Bragg}$ is the total chirp of the grating.

Etching PMMA-POFs using pure acetone or a mixture of acetone and alcohol is a simple technique commonly used by several authors to reduce the diameter of POFs in order to improve the FBG reflectivity [62], the FBG sensitivity to stress, temperature and pressure [111] and provide humidity sensitive or insensitive [2,112] or even to connectorize POFs [113]. Taking into account the easiness and flexibility of the POFs etching process, together with the possibility of writing CFBGs using a single phase mask, through the tapering method, Min et al. reported experimental results on the first CFBG in a tapered BDK-doped PMMA mPOF at the 850 nm region [114]. To achieve a change in the radius along the fiber length, the authors laid the fiber in a container filled with acetone and took advantage of its volatile nature to create a linear taper profile. The 1 cm long grating was inscribed in the tapered region of the mPOF by using a 567.8 nm uniform phase mask, a 248 nm KrF UV laser and by keeping the fiber under 1.0% constant strain. The authors report the fabrication of different taper profiles, showing that a POFBG with 2.6 nm bandwidth could be achieved with a taper slope of 14 μm/cm and by straining the fiber 1.5% after the inscription process. In the same year, the authors reported also the inscription of a CFBG in the same fiber at the 1,550 nm region, but for a taper slope of 10 μm/cm, allowing to achieve a maximum bandwidth after stress release of 4.9 nm, corresponding to a chirp of ~0.5 nm/mm and a dispersion value of 11.2 ps/nm [115]. The dependence of the grating dispersion with strain was also subject of study, revealing positive values ranging between ~11.2 and 513.6 ps/nm for the fiber in rest and with 1.04% strain, respectively, while negative values ranging between 490.6 and 15.9 ps/nm for strain values of 0.94% and 15.9 ps/nm, respectively. The different dispersion signs from 0.94% to 1.04% was associated with the strain in which the grating was inscribed (~1%) [115].

3.8.3 ANNEALING IN HOT WATER

Due to the low T_g of polymer optical fiber materials, their mechanical properties can be manipulated with simple methodologies such as hot water [116], soldering iron [117], and the flame from an alcohol lamp [118]. Annealing process offers the capability to relax the polymer chains frozen in stress after the heat-drawing process. This, in turn, allows to permanently blue-shift the resonance Bragg wavelength. Based on this property, Min et al. [119] reported the fabrication of CFBGs in POFs through a gradient annealing process made in hot water. A BDK-PMMA mPOF was initialed annealed at 55°C in hot water for a period of 10 minutes. Then, a POFBG was written with a length of 1 cm using a uniform phase mask, allowing to get a grating located at the 850 nm region. The fiber was then positioned vertically in a water container at 55°C, with 4 mm of the grating region immersed in water, while the rest was kept at room temperature. The schematic may be seen in Figure 3.18.

The process allowed the authors to create gradient thermal annealing due to the difference in temperatures, and thus, a permanent blue-shift of the left part of the POFBG spectrum was observed. Using this methodology for 8.5 minutes, the authors were able to report a POFBG with 11 nm bandwidth. Posterior characterization of the grating to strain, temperature, and humidity, revealed that the central wavelength is linearly shifted while keeping a stable bandwidth, which was similar to the response of CFBGs fabricated through chirped phase masks.

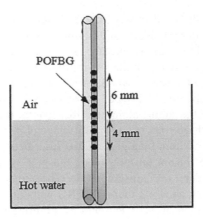

FIGURE 3.18 Schematic of the fabrication of a CFBG based POF in hot water, implemented in Ref. [119].

3.8.4 DIRECT WRITING METHODS

The flexibility and precision on the control of the grating period, using direct writing methods, such as the PbP and LbL, offers another methodology to efficiently write CFBGs in POFs as described in Subsection 2.8.2. Nevertheless, the possibility to change the refractive index with fs lasers, via nonlinear multiphoton absorption, for wavelengths where polymers are transparent, allows the use of large diameter fibers without the need to perform etching procedures. Taking these advantages, Theodosiou et al. [88] showed the possibility to write a CFBG in the most transparent POF material (CYTOP®), using a commercial 500 μm diameter GI-POF and implementing the PbP technology through a fs laser operating at 517 nm [88]. The fiber has a core diameter of 62.5 μm surrounded by a 20 μm cladding and protected by an overcladding that makes up 500 μm total diameter. The number of modes allowed to propagate in this fiber is about 140. In order to excite the fundamental mode of the fiber, the authors wrote the gratings with plane widths of 15 μm to cover the central region of the fiber core where the refractive index peaks. The period of each plane was increased by steps of 7.65 pm for a total of 2,000 periods. The initial period was ~2.3 μm for a 4th order CFBG with a resonance wavelength centered at 1,560 nm. The measured bandwidth was about 10 nm, which translates to a grating chirp of ~2.22 nm/mm [88].

3.9 LONG PERIOD GRATINGS (LPGs) IN POFs

Long-period gratings are fiber structures composed of periodic perturbations of the refractive index along the longitudinal fiber axis. The period of these structures is on the order of hundreds of microns and allows the coupling between the fundamental core mode and co-propagating cladding modes, following Equation 2.6. This coupling will create attenuation dips in the transmission spectrum. A wide range of applications can be found for devices based on this technology including gain equalizers for erbium-doped fiber amplifiers, channel routers in optical add/drop multiplexers, dispersion compensators and several types of sensors [120–122].

Due to the low Young's modulus and biocompatibility of POFs, they could increase the potentialities of LPG in the sensors industry. Hence, different works have been reported regarding the inscription and characterization of LPGs in POFs. The inscription methodologies follow the ones already developed for silica fibers; however, slight modifications have been implemented due to the physical differences between both materials.

3.9.1 MECHANICAL IMPRINTING METHOD

The first report of an LPG in a POF was made in 2004 by van Eijkelenborg and co-workers [123]. The motivation for the work was mainly due to the capability to produce fibers with SM behavior through the use of microstructured designs [124]. The grating was fabricated using the methodology previously demonstrated for the creation of LPGs in silica fibers [125], where the fiber is mechanically pressed using a periodically grooved plate as exemplified in Figure 3.19.

As the grooved plate shown in Figure 3.19 is pressed onto the fiber, a stress-induced periodic refractive index change will occur, leading to the formation of attenuation dips in the transmission spectrum. To mechanically imprint an LPG in a POF, van Eijkelenborg and his colleagues pressed a PMMA rod with 150 triangular grooves with 0.2 mm depth and a period of 1 mm onto a PMMA mPOF [123]. The methodology allowed to control the position of the resonance wavelength, by changing the angle between the fiber and the grooves; control the linewidth of the resonances by increasing the length of the fiber under pressure; and increase the depth of the resonant wavelength by changing the applied pressure. Results measured when the mPOF was subjected to a weight of ~5.9 kg resulted in a grating with a transmission dip at around 530 and 570 nm, presenting 34 dB in strength [123]. The produced gratings also had the capability to be erased after removing the weight from the grooved rod. Two years later, the plate supporting the fiber was heated with temperatures around 60°C–70°C to induce stable LPGs [126]. Gratings produced by this method have shown attenuation dips at the visible region, with strengths up to 18 dB, retaining their features with attenuations dips greater than 8 dB after being stored at 60°C for 2 weeks. An example of a grating inscribed through this methodology may be seen in Figure 3.20.

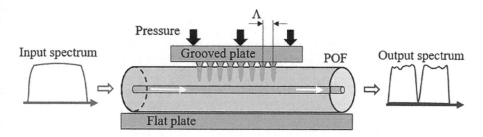

FIGURE 3.19 LPG fabrication through the mechanical imprinting method, using a periodically grooved plate, allowing to observe attenuation dips in the transmission spectrum.

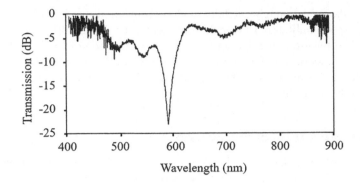

FIGURE 3.20 Transmission spectrum obtained from an LPG written in a SM-mPOF, produced by heating the fiber under mechanical stress. (Structured sold by Kiriama Pty., Ltd.)

3.9.2 Amplitude Mask Technique

Despite the good results produced by mechanical imprinting methods, LPGs in POFs have been essentially fabricated through side irradiation processes using different light sources. The first of those studies was presented in 2004 by Li et al. for a grating imprinted through the amplitude mask technique in a SM-SI photosensitive fiber, using a low-cost mercury lamp [127]. In this technique, the amplitude mask is placed above the fiber. This mask consists of an array of transparent windows that forms an illuminated-shadow periodical pattern over the fiber as shown in Figure 3.21, yielding a grating with the same period of the amplitude mask.

Considering the work reported by Li et al. [127], a mercury lamp with a fluence of $0.3\,\text{mW/cm}^2$ was placed above an amplitude mask with 275 µm pitch. The fiber used was composed of a 176 µm poly(butyl acrylate) (PBA) cladding and 10 µm photosensitive core composed of poly(MMA-co-MVK-co-BzMA). To avoid photosensitivity by the cladding material, the authors filtered part of the UV spectrum with a 1.5 mm Pyrex glass, allowing to remove the wavelengths at which the PBA shows absorption. By irradiating the fiber with periods of 30 seconds in a total exposure

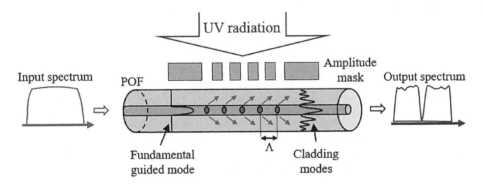

FIGURE 3.21 LPG fabrication through the amplitude mask technique. The irradiated region will modulate the refractive index of the core region with the period of the phase mask.

time of 200 seconds, the authors were able to report a permanent LPG in the core POF located at 1,568 nm, with 3 dB transmission loss and 3 nm bandwidth [127]. The same technique was then used to inscribe LPGs in other types of POFs and with different UV light sources [59–61,128–130].

3.9.3 POINT-BY-POINT TECHNIQUE

The simplicity of the mechanical imprinted method as well as the amplitude mask technique makes these technologies good candidates for the inscription of LPGs. However, they provide low flexibility, since it is necessary a specific grooved plate or amplitude mask to imprint a specific grating period. Instead, the use of PbP or LbL methodologies are more desired, as they allow flexible control over the period, length, and index modulation of the grating. This technique has been first proposed in Ref. [131] and it is nowadays the most used technique. In its simplest form, the process is based on the irradiation of a section of fiber by a thin slit aperture, allowing the refractive index modification at the exposure area. After enough exposure, the laser is switched off and the fiber or the laser beam is moved forward with the desired period, which is then followed by another exposure. The process is repeated up to the desired grating length. A schematic example of the fabrication process of an LPG through the PbP method may be seen in Figure 3.22.

LPGs in POF through the PbP method were first proposed by Saéz-Rodríguez et al. in 2009 [132], five years after the first demonstration of an LPG in a POF through mechanical imprinting method. The inscription was made for an undoped PMMA mPOF using a continuous-wave (CW) 325 nm, 30 mW HeCd laser. The diameter of the UV beam was reduced by focusing it into the core of the fiber through a spherical lens. The grating was written with 1 mm period along a 20 mm length, creating a transmission spectrum with six resonance dips, located in wavelengths between 600 and 1,100 nm and with transmission losses up to 8.5 dB.

Still, in 2009, the first attempt to inscribe LPGs in POFs through the use of focused femtosecond laser pulses was demonstrated in Ref. [133]. The experiment was conducted in a commercial undoped PMMA SM-POF using a 400 nm ultrafast laser with pulse duration of 100 fs and pulse energy of 100 nJ. The grating was

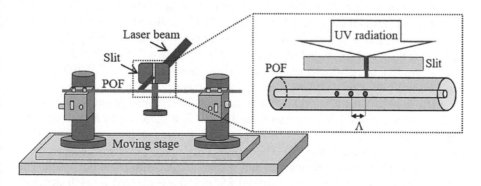

FIGURE 3.22 Inscription of LPGs through the PbP technique.

seen under microscope revealing refractive index changes with ~180 μm period and ~1.8 μm width. The spectral analysis of the grating was however not reported.

The use of photosensitive POFs for the fabrication of gratings is preferred than undoped ones either due to the time needed for the inscription and also due to the possibility to achieve stronger refractive index modifications [16]. Based on these properties, Wang et al. reported the fabrication of a new high photosensitive POF composed of 195 μm PBA cladding material and 11 μm core which was doped with 9-VA for enhanced photosensitivity at the ultraviolet/visible region [17]. An LPG was written on the POF through the PbP technique using a 355 nm solid-state pulsed laser with fluence of 168 mW/cm^2 [17]. The grating was written with 836 μm period, length of 40.1 mm, and with 2 minutes exposure time per point. The result was two transmission dips located at ~1,550 nm, showing losses up to 13 dB and with bandwidths of 14 nm.

In a later work, Kowal et al. [134] used a 300 μm SM-PMMA mPOF with a 40 μm TSB doped overcladding in order to enhance the inscription time under UV radiation. The motivation relied on the shrinkage process of TSB material, which induces a refractive index modification of the exposed area through the stress optic effect. The inscription was made with dual side exposure through the PbP technique with period of 1 mm and using a 30 mW CW HeCd laser operating at 325 nm. After inscribing the grating with 10 periods and with an exposure time of 42 seconds/period, the authors were able to report an LPG located at the 850 nm region and with a transmission dip loss of 20 dB and bandwidth of 45 nm. The result revealed a six-fold improvement compared with the result for the undoped PMMA mPOF (without the TSB doped external layer) [134]. Following this work, the same authors also reported an innovative way to reduce the inscription time of conventional undoped POFs, namely by the diffusion of a mixture of azobenzene and methanol through the cladding region of the POF [135]. By proper diffusing time (i.e., 8 μm thick doped layer obtained for 10 minutes diffusion time), the authors were able to report a stable LPG located at 750 nm with 15 dB depth and 22 nm bandwidth, using a grating pitch of 1.2 mm, exposure time per point of 20 seconds and a total of sixteen periods [135].

The reduction of the gratings inscription time in POFs has been mainly accomplished through the use of dopants [16,134,135]. However, depending on the peak absorption of the fiber material or dopant type, it may be a good opportunity to change the wavelength of the source to accommodate the best performance in terms of grating strength [19] or inscription time [21,73,136]. Oliveira et al. were pioneers in the demonstration the capability to write Bragg gratings in undoped PMMA POFs in few seconds by switching from the most common 325 nm HeCd UV laser to a 248 nm KrF laser, which offered much higher photosensitivity [73]. BDK has its peak absorption at this wavelength, thus, shorter inscription times may be achieved for BDK-doped POFs [21]. Taking into account these considerations, Min et al. [137] used the 248 nm UV source to write an LPG in a BDK-doped mPOF with the PbP technique with an inscription time per writing point of 2 seconds (lower than the 20 seconds reported in [135]). The laser energy and repetition rate were set to ~3 mJ and 1 Hz, respectively, for a 25 mm grating length with period of 1 mm. The obtained gratings were located at around 850 nm and presented a notch depth of about 20 dB.

LPGs in POFs have essentially been created using UV sources [17,59,134,135, 137,60,61,127–129,131–133] due to the higher photosensitivity and absorption of polymers at short wavelengths. Regarding the absorption effect, it is known that upon UV absorption the temperature of the irradiated region is increased, leading to the occurrence of a material shrinkage process and inherently to a refractive index modification of the material due to the stress optic effect [134]. On the other hand, CO_2 lasers (operating wavelength 10.6 μm) have been the preferred choice for the fabrication of LPGs in silica fibers compared to UV sources. The reason is essentially related to their lower cost and also because they do not require the use of photosensitive fibers [138].

The advantages of CO_2 laser inscription in silica fibers also apply to POFs, and thus, they have also been the subject of study [139,140]. Preliminary works reporting the use of CO_2 laser irradiation in a POF may be found in 2010 for a fiber that had been deformed with a period of 588 μm [139]. Further details were not discussed in these preliminary results. Later in 2013, authors from another group reported the use of a CO_2 laser to inscribe a fiber Bragg grating in the THz region on a 400 μm low-density PS fiber through the PbP technique and with laser output power of 1.5 W [141]. The structure was periodically engraved, showing notches with 110 μm deep and ~170 μm wide, which were probably originated from the high-thermal absorption of the polymer material at this wavelength region. The grating was seen in transmission having 15 dB transmission loss, 4 GHz bandwidth, and centered at 375 GHz.

The use of a CO_2 laser for the inscription of an LPG at the visible region was reported three years later by Bundalo et al. [140]. For that, the authors used a commercial CO_2 laser (DC-K40III, Liaocheng Shenhui Laser Equipment Co.) and operated it at the minimum power. The inscription was made PbP with 0.8 and 1 mm periods in a 310 μm SM PMMA mPOF, allowing to obtain gratings with extinction of 25 dB, located at the low loss region of PMMA, at around (590–670 nm) [140]. Periodic uniform grooves were observed under a microscope at the surface of the POF and were associated with the ablation of the polymer material due to the rise of temperatures upon laser incidence.

REFERENCES

1. G. D. Peng, Z. Xiong, and P. L. Chu, "Photosensitivity and gratings in dye-doped polymer optical fibers," *Opt. Fiber Technol.*, vol. 5, no. 2, pp. 242–251, 1999.
2. R. Oliveira, T. H. R. Marques, L. Bilro, R. Nogueira, and C. M. B. Cordeiro, "Multiparameter POF sensing based on multimode interference and fiber Bragg grating," *J. Light. Technol.*, vol. 35, no. 1, pp. 3–9, 2017.
3. H. Dobb, D. J. Webb, K. Kalli, A. Argyros, M. C. J. Large, and M. A. van Eijkelenborg, "Continuous wave ultraviolet light-induced fiber Bragg gratings in few - and single-mode microstructured polymer optical fibers," *Opt. Lett.*, vol. 30, no. 24, pp. 3296–3298, 2005.
4. M. Koerdt et al., "Fabrication and characterization of Bragg gratings in a graded-index perfluorinated polymer optical fiber," in *2nd International Conference on System-Integrated Intelligence: Challenges for Product and Production Engineering*, 2014, vol. 15, pp. 138–146.

5. A. Lacraz, M. Polis, A. Theodosiou, C. Koutsides, and K. Kalli, "Femtosecond laser inscribed Bragg gratings in low loss CYTOP polymer optical fiber," *IEEE Photonics Technol. Lett.*, vol. 27, no. 7, pp. 693–696, 2015.
6. R. Oliveira et al., "Bragg gratings inscription in highly birefringent microstructured POFs," *IEEE Photonics Technol. Lett.*, vol. 28, no. 6, pp. 621–624, 2016.
7. I. P. Johnson et al., "Optical fibre Bragg grating recorded in TOPAS cyclic olefin copolymer," *Electron. Lett.*, vol. 47, no. 4, pp. 271–272, 2011.
8. A. Fasano et al., "Fabrication and characterization of polycarbonate microstructured polymer optical fibers for high-temperature-resistant fiber Bragg grating strain sensors," *Opt. Mater. Express*, vol. 6, no. 2, pp. 649–659, 2016.
9. G. Woyessa, A. Fasano, C. Markos, A. Stefani, H. K. Rasmussen, and O. Bang, "Zeonex microstructured polymer optical fiber: Fabrication friendly fibers for high temperature and humidity insensitive Bragg grating sensing," *Opt. Mater. Express*, vol. 7, no. 1, pp. 286–295, 2017.
10. G. N. Harbach, H. G. Limberger, and R. P. Salathé, "Influence of humidity and temperature on polymer optical fiber Bragg gratings," in *Advanced Photonics & Renewable Energy*, 2010, p. BTuB2.
11. W. Yuan et al., "Humidity insensitive TOPAS polymer fiber Bragg grating sensor," *Opt. Express*, vol. 19, no. 20, pp. 19731–19739, 2011.
12. C. Markos, A. Stefani, K. Nielsen, H. K. Rasmussen, W. Yuan, and O. Bang, "High-Tg TOPAS microstructured polymer optical fiber for fiber Bragg grating strain sensing at 110 degrees," *Opt. Express*, vol. 21, no. 4, pp. 4758–4765, 2013.
13. G. Emiliyanov et al., "Localized biosensing with Topas microstructured polymer optical fiber," *Opt. Lett.*, vol. 32, no. 5, pp. 460–462, 2007.
14. A. Leal-Junior et al., "Characterization of a new polymer optical fiber with enhanced sensing capabilities using a Bragg grating," *Opt. Lett.*, vol. 43, no. 19, pp. 4799–4802, 2018.
15. J. Yu, X. Tao, and H. Tam, "Trans-4-stilbenemethanol-doped photosensitive polymer fibers and gratings," *Opt. Lett.*, vol. 29, no. 2, pp. 156–158, 2004.
16. Y. Luo, Q. Zhang, H. Liu, and G.-D. Peng, "Gratings fabrication in benzildimethylketal doped photosensitive polymer optical fibers using 355 nm nanosecond pulsed laser," *Opt. Lett.*, vol. 35, no. 5, pp. 751–753, 2010.
17. T. Wang et al., "Enhancing photosensitivity in near UV/vis band by doping 9-vinylanthracene in polymer optical fiber," *Opt. Commun.*, vol. 307, pp. 5–8, 2013.
18. J. Bonefacino et al., "Ultra-fast polymer optical fibre Bragg grating inscription for medical devices," *Light Sci. Appl.*, vol. 7, no. 3, p. 17161, 2018.
19. D. Sáez-Rodríguez, K. Nielsen, H. K. Rasmussen, O. Bang, and D. J. Webb, "Highly photosensitive polymethyl methacrylate microstructured polymer optical fiber with doped core," *Opt. Lett.*, vol. 38, no. 19, pp. 3769–3772, 2013.
20. H. Liu, G. Peng, and P. Chu, "Polymer fiber Bragg gratings with 28-dB transmission rejection," *IEEE Photonics Technol. Lett.*, vol. 14, no. 7, pp. 935–937, 2002.
21. A. Pospori, C. A. F. Marques, O. Bang, D. J. Webb, and P. André, "Polymer optical fiber Bragg grating inscription with a single UV laser pulse," *Opt. Express*, vol. 25, no. 8, pp. 9028–9038, 2017.
22. Z. F. Zhang, C. Zhang, X. M. Tao, G. F. Wang, and G. D. Peng, "Inscription of polymer optical fiber Bragg grating at 962 nm and its potential in strain sensing," *IEEE Photonics Technol. Lett.*, vol. 22, no. 21, pp. 1562–1564, 2010.
23. A. Stefani, C. Markos, and O. Bang, "Narrow bandwidth 850-nm fiber Bragg gratings in few-mode polymer optical fibers," *IEEE Photonics Technol. Lett.*, vol. 23, no. 10, pp. 660–662, 2011.

24. G. Statkiewicz-Barabach, K. Tarnowski, D. Kowal, P. Mergo, and W. Urbanczyk, "Fabrication of multiple Bragg gratings in microstructured polymer fibers using a phase mask with several diffraction orders," *Opt. Express*, vol. 21, no. 7, pp. 8521–8534, 2013.

25. C. A. F. Marques, L. Bilro, N. J. Alberto, D. J. Webb, and R. Nogueira, "Narrow bandwidth Bragg gratings imprinted in polymer optical fibers for different spectral windows," *Opt. Commun.*, vol. 307, pp. 57–61, 2013.

26. R. Oliveira, L. Bilro, J. Heidarialamdarloo, and R. Nogueira, "Fast inscription of Bragg grating arrays in undoped PMMA mPOF," in *24th International Conference on Optical Fiber Sensors*, H. J. Kalinowski, J. L. Fabris and W. J. Bock, Eds. Bellingham: SPIE, 2015, vol. 9634, p. 96344X.

27. W. Yuan, A. Stefani, and O. Bang, "Tunable polymer fiber Bragg grating (FBG) inscription: Fabrication of dual-FBG temperature compensated polymer optical fiber strain sensors," *Photonics Technol. Lett.*, vol. 24, no. 5, pp. 401–403, 2012.

28. G. Rajan, M. Y. Mohd Noor, E. Ambikairajah, and G.-D. Peng, "Inscription of multiple Bragg gratings in a singlemode polymer optical fiber using a single phase mask and its analysis," *IEEE Sens. J.*, vol. 14, no. 7, pp. 2384–2388, 2014.

29. I. P. Johnson, D. J. Webb, K. Kalli, M. C. Large, and A. Argyros, "Multiplexed FBG sensor recorded in multimode microstructured polymer optical fibre," in *Photonic Crystal Fibres IV*, K. Kalli and W. Urbanczyk, Eds. Bellingham: SPIE, 2010, vol. 7714, p. 77140D.

30. A. Pospori, C. A. F. Marques, G. Sagias, H. Lamela-Rivera, and D. J. Webb, "Novel thermal annealing methodology for permanent tuning polymer optical fiber Bragg gratings to longer wavelengths," *Opt. Express*, vol. 26, no. 2, pp. 1411–1421, 2018.

31. A. Fasano, G. Woyessa, J. Janting, H. K. Rasmussen, and O. Bang, "Solution-mediated annealing of polymer optical fiber Bragg gratings at room temperature," *IEEE Photonics Technol. Lett.*, vol. 29, no. 8, pp. 687–690, 2017.

32. X. Hu, C.-F. J. Pun, H.-Y. Tam, P. Mégret, and C. Caucheteur, "Tilted Bragg gratings in step-index polymer optical fiber," *Opt. Lett.*, vol. 39, no. 24, pp. 6835–6838, 2014.

33. C. A. F. Marques, P. Antunes, P. Mergo, D. J. Webb, and P. André, "Chirped Bragg gratings in PMMA step-index polymer optical fiber," *IEEE Photonics Technol. Lett.*, vol. 29, no. 6, pp. 500–503, 2017.

34. L. Pereira et al., "Phase-shifted Bragg grating inscription in PMMA microstructured POF using 248 nm UV radiation," *J. Light. Technol.*, vol. 35, no. 23, pp. 5176–5184, 2017.

35. G. Statkiewicz-Barabach, P. Mergo, and W. Urbanczyk, "Bragg grating-based Fabry–Perot interferometer fabricated in a polymer fiber for sensing with improved resolution," *J. Opt.*, vol. 19, no. 1, p. 015609, 2017.

36. A. Lacraz, A. Theodosiou, and K. Kalli, "Femtosecond laser inscribed Bragg grating arrays in long lengths of polymer optical fibres; a route to practical sensing with POF," *Electron. Lett.*, vol. 52, no. 19, pp. 1626–1627, 2016.

37. A. Theodosiou, A. Lacraz, A. Stassis, M. Komodromos, and K. Kalli, "Plane-by-Plane femtosecond laser inscription method for single-peak Bragg gratings in multimode CYTOP polymer optical fibre," *J. Light. Technol.*, vol. 35, no. 24, pp. 5404–5410, 2017.

38. A. Othonos, "Fiber Bragg gratings," *Rev. Sci. Instrum.*, vol. 68, no. 12, pp. 4309–4341, 1997.

39. Q. Zhang, D. A. Brown, L. Reinhart, T. F. Morse, J. Q. Wang, and G. Xiao, "Tuning Bragg wavelength by writing gratings on prestrained fibers," *IEEE Photonics J.*, vol. 6, no. 7, pp. 839–841, 1994.

40. G. Woyessa, K. Nielsen, A. Stefani, C. Markos, and O. Bang, "Temperature insensitive hysteresis free highly sensitive polymer optical fiber Bragg grating humidity sensor," *Opt. Express*, vol. 24, no. 2, pp. 1206–1213, 2016.

41. M. Koerdt et al., "Fabrication and characterization of Bragg gratings in perfluorinated polymer optical fibers and their embedding in composites," *Mechatronics*, vol. 34, pp. 137–146, 2016.

42. Z. F. Zhang and X. Ming Tao, "Synergetic effects of humidity and temperature on PMMA based fiber Bragg gratings," *J. Light. Technol.*, vol. 30, no. 6, pp. 841–845, 2012.

43. D. Sáez-Rodríguez, K. Nielsen, O. Bang, and B. Ortega, "Compact multichannel demultiplexer for WDM-POF networks based on spatially overlapped FBGs," *Electron. Lett.*, vol. 52, no. 8, pp. 635–637, 2016.

44. C. Jiang, M. G. Kuzyk, J.-L. Ding, W. E. Johns, and D. J. Welker, "Fabrication and mechanical behavior of dye-doped polymer optical fiber," *J. Appl. Phys.*, vol. 92, no. 1, pp. 4–12, 2002.

45. W. Yuan et al., "Improved thermal and strain performance of annealed polymer optical fiber Bragg gratings," *Opt. Commun.*, vol. 284, no. 1, pp. 176–182, 2011.

46. K. E. Carroll, C. Zhang, D. J. Webb, K. Kalli, A. Argyros, and M. C. Large, "Thermal response of Bragg gratings in PMMA microstructured optical fibers," *Opt. Express*, vol. 15, no. 14, pp. 8844–8850, 2007.

47. P. Stajanca, O. Cetinkaya, M. Schukar, P. Mergo, D. J. Webb, and K. Krebber, "Molecular alignment relaxation in polymer optical fibers for sensing applications," *Opt. Fiber Technol.*, vol. 28, pp. 11–17, 2016.

48. G. Woyessa et al., "Zeonex-PMMA microstructured polymer optical FBGs for simultaneous humidity and temperature sensing," *Opt. Lett.*, vol. 42, no. 6, pp. 1161–1164, 2017.

49. A. Abang and D. J. Webb, "Effects of annealing, pre-tension and mounting on the hysteresis of polymer strain sensors," *Meas. Sci. Technol.*, vol. 25, no. 1, p. 015102, 2014.

50. A. Pospori et al., "Annealing effects on strain and stress sensitivity of polymer optical fibre based sensors," in *Micro-Structured and Specialty Optical Fibres IV*, K. Kalli and A. Mendez, Eds. Bellingham: SPIE, 2016, vol. 9886, p. 98860V.

51. H. Levine and L. Slade, "Water as a plasticizer: physico-chemical aspects of low-moisture polymeric systems," in *Water Science Reviews 3: Water Dynamics*, F. Franks, Ed. Cambridge: Cambridge University Press, 1988, pp. 79–185.

52. L. S. A. Smith and V. Schmitz, "The effect of water on the glass transition temperature of poly (methyl methacrylate)," *Polymer*, vol. 29, no. 10, pp. 1871–1878, 1988.

53. D. R. G. Williams, P. E. M. Allen, and V. T. Truong, "Glass transition temperature and stress relaxation of methanol equilibrated poly (methyl methacrylate)," *Eur. Polym. J.*, vol. 22, no. 11, pp. 911–919, 1986.

54. S. Kiesel, K. Peters, T. Hassan, and M. Kowalsky, "Behaviour of intrinsic polymer optical fibre sensor for large-strain applications," *Meas. Sci. Technol.*, vol. 18, no. 10, pp. 3144–3154, 2007.

55. P. Antunes, H. Lima, J. Monteiro, and P. S. Andre, "Elastic constant measurement for standard and photosensitive single mode optical fibres," *Microw. Opt. Technol. Lett.*, vol. 50, no. 9, pp. 2467–2469, 2008.

56. X. Xingsheng et al., "Birefringent gratings induced by polarized laser in azobenzene-doped Poly(methyl methccrylate) optical fibers," in *Fiber Optics and Optoelectronics for Network Applications*, J. Liu and Z. Wang, Eds. Bellingham: SPIE, 2001, vol. 4603, pp. 260–265.

57. X. Xu, H. Ming, Q. Zhang, and Y. Zhang, "Properties of Raman spectra and laser-induced birefringence in polymethyl methacrylate optical fibres," *J. Opt. A Pure Appl. Opt.*, vol. 4, no. 3, pp. 237–242, 2002.

58. X. Xingsheng, M. Hai, and Z. Qijin, "Properties of polarized laser-induced birefringent gratings in azobenzene-doped poly (methyl methecrylate) optical fibers," *Opt. Commun.*, vol. 204, pp. 137–143, 2002.

59. Y. Luo et al., "Birefringent azopolymer long period fiber gratings induced by 532 nm polarized laser," *Opt. Commun.*, vol. 282, no. 12, pp. 2348–2353, 2009.

60. Z. Li, H. Ma, Q. Zhang, and H. Ming, "Birefringence grating within a single mode polymer optical fibre with photosensitive core of azobenzene copolymer," *J. Optoelectron. Adv. Mater.*, vol. 7, no. 2, pp. 1039–1046, 2005.
61. M. Hui et al., "Analysis of photosensitivity of copolymer optical fibre preform," *Chinese Phys. Lett.*, vol. 21, no. 11, pp. 2252–2254, 2004.
62. X. Hu, C.-F. J. Pun, H.-Y. Tam, P. Mégret, and C. Caucheteur, "Highly reflective Bragg gratings in slightly etched step-index polymer optical fiber," *Opt. Express*, vol. 22, no. 15, pp. 18807–18817, 2014.
63. C. Caucheteur et al., "Transverse strain measurements using the birefringence effect in fiber Bragg gratings," *IEEE Photonics Technol. Lett.*, vol. 19, no. 13, pp. 966–968, 2007.
64. X. Hu et al., "Polarization effects in polymer FBGs: Study and use for transverse force sensing," *Opt. Express*, vol. 23, no. 4, pp. 4581–4590, 2015.
65. Y. Zhang et al., "Casting preforms for microstructured polymer optical fibre fabrication," *Opt. Express*, vol. 14, no. 12, pp. 5541–5547, 2006.
66. J. Olszewski, P. Mergo, K. Gąsior, and W. Urbańczyk, "Highly birefringent microstructured polymer fibers optimized for a preform drilling fabrication method," *J. Opt.*, vol. 15, no. 075713, 2013.
67. H. Ebendorff-Heidepriem, T. M. Monro, M. A. van Eijkelenborg, and M. C. J. Large, "Extruded high-NA microstructured polymer optical fibre," *Opt. Commun.*, vol. 273, no. 1, pp. 133–137, 2007.
68. K. Cook et al., "Air-structured optical fiber drawn from a 3D-printed preform," *Opt. Lett.*, vol. 40, no. 17, pp. 3966–3969, 2015.
69. N. A. Issa, M. A. van Eijkelenborg, M. Fellew, F. Cox, G. Henry, and M. C. J. Large, "Fabrication and study of microstructured optical fibers with elliptical holes," *Opt. Lett.*, vol. 29, no. 12, pp. 1336–1338, 2004.
70. M. K. Szczurowski, T. Martynkien, G. Statkiewicz-Barabach, W. Urbanczyk, and D. J. Webb, "Measurements of polarimetric sensitivity to hydrostatic pressure, strain and temperature in birefringent dual-core microstructured polymer fiber," *Opt. Express*, vol. 18, no. 12, pp. 12076–12087, 2010.
71. T. Martynkien, P. Mergo, and W. Urbanczyk, "Sensitivity of birefringent microstructured polymer optical fiber to hydrostatic pressure," *IEEE Photonics Technol. Lett.*, vol. 25, no. 16, pp. 1562–1565, 2013.
72. I. Abe, H. J. Kalinowski, O. Frazão, J. L. Santos, R. N. Nogueira, and J. L. Pinto, "Superimposed Bragg gratings in high-birefringence fibre optics: Three-parameter simultaneous measurements," *Meas. Sci. Technol.*, vol. 15, no. 8, pp. 1453–1457, 2004.
73. R. Oliveira, L. Bilro, and R. Nogueira, "Bragg gratings in a few mode microstructured polymer optical fiber in less than 30 seconds," *Opt. Express*, vol. 23, no. 8, pp. 10181–10187, 2015.
74. T. Erdogan and J. E. Sipe, "Tilted fiber phase gratings," *J. Opt. Soc. Am. A*, vol. 13, no. 2, pp. 296–313, 1996.
75. Y. X. Jin, C. C. Chan, X. Y. Dong, and Y. F. Zhang, "Temperature-independent bending sensor with tilted fiber Bragg grating interacting with multimode fiber," *Opt. Commun.*, vol. 282, no. 19, pp. 3905–3907, 2009.
76. N. J. Alberto, C. A. F. Marques, J. L. Pinto, and R. N. Nogueira, "Three-parameter optical fiber sensor based on a tilted fiber Bragg grating," *Appl. Opt.*, vol. 49, no. 31, pp. 6085–6091, 2010.
77. T. Guo, F. Liu, B. Guan, and J. Albert, "Tilted fiber grating mechanical and biochemical sensors," *Opt. Laser Technol.*, vol. 78, pp. 19–33, 2016.
78. C. Caucheteur, T. Guo, and J. Albert, "Review of plasmonic fiber optic biochemical sensors: Improving the limit of detection," *Anal. Bioanal. Chem.*, vol. 407, no. 14, pp. 3883–3897, 2015.

79. N. Cennamo et al., "An easy way to realize SPR aptasensor: A multimode plastic optical fiber platform for cancer biomarkers detection," *Talanta*, vol. 140, pp. 88–95, 2015.

80. X. Hu, P. Mégret, and C. Caucheteur, "Surface plasmon excitation at near-infrared wavelengths in polymer optical fibers," *Opt. Lett.*, vol. 40, no. 17, pp. 3998–4001, 2015.

81. D. R. Huber, "1.5 µm narrow bandwidth in-fiber gratings," in *Proc. of IEEE LEOS Annual Meeting (LEOS' 91)*, 1991, p. Paper OE3.1.

82. C. Broadway et al., "Microstructured polymer optical fibre sensors for opto-acoustic endoscopy," in *Micro-Structured and Specialty Optical Fibres IV*, K. Kalli and A. Mendez, Eds. Bellingham: SPIE, 2016, vol. 9886, p. 98860S.

83. C. Broadway et al., "Fabry-Perot micro-structured polymer optical fibre sensors for opto-acoustic endoscopy," in *Biophotonics South America*, C. Kurachi, K. Svanberg, B. J. Tromberg and V. S. Bagnato, Eds. Bellingham: SPIE, 2015, vol. 9531, p. 953116.

84. D. Webb et al., "Grating and interferometric devices in POF," in *14th International Conference on Polymer Optical Fibers*, 2005, pp. 325–328.

85. H. Dobb et al., "Grating based devices in polymer optical fibre," in *Optical Sensing II*, B. Culshaw, A. G. Mignani, H. B. Bartelt and L. R. Jaroszewicz, Eds. Bellingham: SPIE, 2006, vol. 6189, p. 618901.

86. A. Pospori and D. J. Webb, "Performance analysis of polymer optical fibre based Fabry-Perot sensor formed by two uniform Bragg gratings," in *Micro-Structured and Specialty Optical Fibres IV*, K. Kalli and A. Mendez, Eds. Bellingham: SPIE, , 2016, vol. 9886, p. 98861F.

87. G. Statkiewicz-Barabach, P. Maniewski, P. Mergo, and W. Urbanczyk, "Fiber Bragg grating-based fabry-Perot interferometer in polymer fiber," in *25th International Conference on Plastic Optical Fibres*, 2016, p. OP22.

88. A. Theodosiou, X. Hu, C. Caucheteur, and K. Kalli, "Bragg Gratings and Fabry-Perot cavities in low-loss multimode CYTOP polymer fiber," *IEEE Photonics Technol. Lett.*, vol. 30, no. 9, pp. 857–860, 2018.

89. H. A. Haus and Y. Lai, "Theory of cascaded quarter wave shifted distributed feedback resonators," *IEEE J. Quantum Electron.*, vol. 28, no. 1, pp. 205–213, 1992.

90. X. Han and J. Yao, "Bandstop-to-bandpass microwave photonic filter using a phase-shifted fiber bragg grating," *J. Light. Technol.*, vol. 33, no. 24, pp. 5133–5139, 2015.

91. O. Xu, J. Zhang, H. Deng, and J. Yao, "Dual-frequency optoelectronic oscillator for thermal-insensitive interrogation of a FBG strain sensor," *IEEE Photonics Technol. Lett.*, vol. 29, no. 4, pp. 357–360, 2017.

92. S. F. Zhou, L. Reekie, Y. T. Chow, H. P. Chan, and K. M. Luk, "Phase-shifted fiber Bragg gratings for terahertz range," *IEEE Photonics Technol. Lett.*, vol. 24, no. 20, pp. 1875–1877, 2012.

93. B. Ortega, R. Min, D. Sáez-Rodríguez, Y. Mi, K. Nielsen, and O. Bang, "Bandpass transmission filters based on phase shifted fiber Bragg gratings in microstructured polymer optical fibers," in *Micro-structured and Specialty Optical Fibres V*, K. Kalli, J. Kanka, A. Mendez and P. Peterka, Eds. Bellingham: SPIE, 2017, vol. 10232, p. 1023209.

94. S. F. Zhou, L. Reekie, H. P. Chan, Y. T. Chow, P. S. Chung, and K. M. Luk, "Characterization and modeling of Bragg gratings written in polymer fiber for use as filters in the THz region," *Opt. Express*, vol. 20, no. 9, pp. 9564–9571, 2012.

95. R. Kashyap, *Fiber Bragg Gratings*. San Diego, CA: Academic Press, 1999.

96. R. Min, C. A. F. Marques, O. Bang, and B. Ortega, "Moiré phase-shifted fiber Bragg gratings in polymer optical fibers," *Opt. Fiber Technol.*, vol. 41, pp. 78–81, 2018.

97. K. O. Hill et al., "Chirped in-fiber Bragg gratings for compensation of optical-fiber dispersion," *Opt. Lett.*, vol. 19, no. 17, pp. 1314–1316, 1994.

98. D. Tosi, "Review of chirped fiber Bragg grating (CFBG) fiber-optic sensors and their applications," *Sensors*, vol. 18, no. 7, p. 2147, 2018.

99. S. Korganbayev et al., "Detection of thermal gradients through fiber-optic Chirped Fiber Bragg Grating (CFBG): Medical thermal ablation scenario," *Opt. Fiber Technol.*, vol. 41, pp. 48–55, 2018.

100. S. Yashiro, T. Okabe, N. Toyama, and N. Takeda, "Monitoring damage in holed CFRP laminates using embedded chirped FBG sensors," *Int. J. Solids Struct.*, vol. 44, no. 2, pp. 603–613, 2007.

101. R. Kashyap, P. F. McKee, R. J. Campbell, and D. L. Williams, "Novel method of producing all fibre photoinduced chirped gratings," *Electron. Lett.*, vol. 30, no. 12, pp. 996–998, 1994.

102. H. Storøy, H. E. Engan, B. Sahlgren, and R. Stubbe, "Position weighting of fiber Bragg gratings for bandpass filtering," *Opt. Lett.*, vol. 22, no. 11, pp. 784–786, 1997.

103. M. C. Farries, K. Sugden, D. C. J. Reid, I. Bennion, A. Molony, and M. J. Goodwin, "Very broad reflection bandwidth (44 nm) chirped fibre gratings and narrow bandpass filters produced by the use of an amplitude mask," *Electron. Lett.*, vol. 30, no. 11, pp. 891–892, 1994.

104. K. Sugden, I. Bennion, A. Molony, and N. J. Copner, "Chirped grating produced in photosensitive optical fibers by fiber deformation during exposure," *Electron. Lett.*, vol. 30, no. 5, pp. 440–442, 1994.

105. J. Martin, J. Lauzon, S. Thibault, and F. Ouellette, "Novel writing technique of long and highly reflective in-fiber Bragg gratings and investigation of the linearly chirped component," in *Optical Fiber Communication Conference*, 1994, p. PD29.

106. L. Dong, J. L. Cruz, L. Reekie, and J. A. Tucknott, "Fabrication of chirped fibre gratings using etched tapers," *Electron. Lett.*, vol. 31, no. 11, pp. 908–909, 1995.

107. C. Voigtlaender, J. Thomas, E. Wikszak, P. Dannberg, S. Nolte, and A. Tünnermann, "Chirped fiber Bragg gratings written with ultrashort pulses and a tunable phase mask," *Opt. Lett.*, vol. 34, no. 12, pp. 1888–1890, 2009.

108. H. Liu, *Polymer Optical Fiber Bragg Gratings*. Kensington: The University of New South Wales, 2003.

109. H. Liu, H. Liu, G. Peng, and T. W. Whitbread, "Tunable dispersion using linearly chirped polymer optical fiber Bragg gratings with fixed center wavelength," *IEEE Photonics Technol. Lett.*, vol. 17, no. 2, pp. 411–413, 2005.

110. J. L. Cruz, L. Dong, S. Barcelos, and L. Reekie, "Fiber Bragg gratings with various chirp profiles made in etched tapers," *Appl. Opt.*, vol. 35, no. 34, pp. 6781–6787, 1996.

111. K. Bhowmik et al., "Intrinsic high-sensitivity sensors based on etched single-mode polymer optical fibers," *IEEE Photonics Technol. Lett.*, vol. 27, no. 6, pp. 604–607, 2015.

112. G. Woyessa, J. K. M. Pedersen, A. Fasano, K. Nielsen, C. Markos, H. K. Rasmussen, and O. Bang, "Simultaneous measurement of temperature and humidity with microstructured polymer optical fiber Bragg gratings," in *25th International Conference on Optical Fiber Sensors*, Y. Chung et al., Eds. Bellingham: SPIE, 2017, vol. 10323, p. 103234.

113. A. Abang and D. J. Webb, "Demountable connection for polymer optical fiber grating sensors," *Opt. Eng.*, vol. 51, no. 8, p. 080503, 2012.

114. R. Min, B. Ortega, and C. A. F. Marques, "Fabrication of tunable chirped mPOF Bragg gratings using a uniform phase mask," *Opt. Express*, vol. 26, no. 4, pp. 4411–4420, 2018.

115. R. Min et al., "Largely tunable dispersion chirped polymer FBG," *Opt. Lett.*, vol. 43, no. 20, pp. 5106–5109, 2018.

116. Y. Mizuno, H. Ujihara, H. Lee, N. Hayashi, and K. Nakamura, "Polymer optical fiber tapering using hot water," *Appl. Phys. Express*, vol. 10, no. 6, p. 062502, 2017.

117. A. A. Jasim et al., "Refractive index and strain sensing using inline Mach–Zehnder interferometer comprising perfluorinated graded-index plastic optical fiber," *Sensors Actuators A Phys.*, vol. 219, pp. 94–99, 2014.

118. F. De-Jun, L. Guan-Xiu, L. Xi-Lu, J. Ming-Shun, and S. Qing-Mei, "Refractive index sensor based on plastic optical fiber with tapered structure," *Appl. Opt.*, vol. 53, no. 10, pp. 2007–2011, 2014.

119. R. Min et al., "Hot water-assisted fabrication of chirped polymer optical fiber Bragg gratings," *Opt. Express*, vol. 26, no. 26, pp. 34655–34664, 2018.

120. S. James and R. Tatam, "Optical fibre long-period grating sensors: Characteristics and application," *Meas. Sci. Technol.*, vol. 14, pp. R49–R61, 2003.

121. Y. Zhu, C. Lu, B. M. Lacquet, P. L. Swart, and S. J. Spammer, "Wavelength-tunable add/drop multiplexer for dense wavelength division multiplexing using long-period gratings and fiber stretchers," *Opt. Commun.*, vol. 208, no. 4–6, pp. 337–344, 2006.

122. D. B. Stegall and T. Erdogan, "Dispersion control with use of long-period fiber gratings," *J. Opt. Soc. Am. A*, vol. 17, no. 2, pp. 304–312, 2000.

123. M. A. van Eijkelenborg, W. Padden, and J. A. Besley, "Mechanically induced long-period gratings in microstructured polymer fibre," *Opt. Commun.*, vol. 236, no. 1–3, pp. 75–78, 2004.

124. M. van Eijkelenborg et al., "Microstructured polymer optical fibre," *Opt. Express*, vol. 9, no. 7, pp. 319–327, 2001.

125. J. H. Lim, K. S. Lee, J. C. Kim, and B. H. Lee, "Tunable fiber gratings fabricated in photonic crystal fiber by use of mechanical pressure," *Opt. Lett.*, vol. 29, no. 4, pp. 331–333, 2004.

126. M. P. Hiscocks, M. A. van Eijkelenborg, A. Argyros, and M. C. J. Large, "Stable imprinting of long-period gratings in microstructured polymer optical fibre," *Opt. Express*, vol. 14, no. 11, pp. 4644–4649, 2006.

127. Z. Li, H. Y. Tam, L. Xu, and Q. Zhang, "Fabrication of long-period gratings in poly (methyl methacrylate-co-methyl vinyl ketone-co-benzyl methacrylate)-core polymer optical fiber by use of a mercury lamp," *Opt. Lett.*, vol. 30, no. 10, pp. 1117–1119, 2005.

128. R. Zheng, Y. Lu, Z. Xie, J. Tao, K. Lin, and H. Ming, "Surface plasmon resonance sensors based on polymer optical fiber," in *1st Asia-Pacific Optical Fiber Sensors Conference*, 2008. doi: 10.1109/APOS.2008.5226283

129. X. Sun, X. Tao, and Z. Zhang, "The investigation on photosensitivity of polymer optical fiber," in *Symposium on Photonics and Optoelectronics (SOPO)*, 2011. doi: 10.1109/SOPO.2011.5780487

130. X. Tao, J. Yu, M. Suleyman Demokan, H. Tam, and D. Yang, "Photosensitivity and grating development in trans-4-stilbenemethanol-doped poly (methyl methacrylate) materials," *Opt. Commun.*, vol. 265, pp. 132–139, 2006.

131. K. O. Hill, B. Malo, K. A. Vineberg, F. Bilodeau, D. C. Johnson, and I. Skinner, "Efficient mode conversion in telecommunication fibre using externally written gratings," *Electron. Lett.*, vol. 26, no. 16, pp. 1270–1272, 1990.

132. D. Sáez-Rodríguez, J. L. C. Munoz, I. Johnson, D. J. Webb, M. C. J. Large, and A. Argyros, "Long period fibre gratings photoinscribed in a microstructured polymer optical fibre by UV radiation," in *Photonic Crystal Fibers III*, K. Kalli, Ed. Bellingham: SPIE, 2009, vol. 7357, p. 73570L.

133. S. J. Liang, P. J. Scully, J. Schille, J. Vaughan, and W. Perrie, "Femtosecond laser induced refractive index structures in polymer optical fibre (POF) for sensing," in *20th International Conference on Optical Fibre Sensors*, J. Jones et al., Eds. Bellingham: SPIE, 2009, vol. 7503, p. 75036S.

134. D. Kowal, G. Statkiewicz-Barabach, P. Mergo, and W. Urbanczyk, "Microstructured polymer optical fiber for long period gratings fabrication using an ultraviolet laser beam," *Opt. Lett.*, vol. 39, no. 8, pp. 2242–2245, 2014.

135. D. Kowal, G. Statkiewicz-Barabach, P. Mergo, and W. Urbanczyk, "Inscription of long period gratings using an ultraviolet laser beam in the diffusion-doped microstructured polymer optical fiber," *Appl. Opt.*, vol. 54, no. 20, pp. 6327–6333, 2015.
136. R. Nogueira, R. Oliveira, L. Bilro, and J. Heidarialamdarloo, "New advances in polymer fiber Bragg gratings," *Opt. Laser Technol.*, vol. 78, pp. 104–109, 2016.
137. R. Min, C. A. F. Marques, K. Nielsen, O. Bang, and B. Ortega, "Fast inscription of long period gratings in microstructured polymer optical fibers," *IEEE Sens. J.*, vol. 18, no. 5, pp. 1919–1923, 2018.
138. Y. Wang, "Review of long period fiber gratings written by CO_2 laser," *J. Appl. Phys.*, vol. 108, no. 8, p. 081101, 2010.
139. H. Y. Tam, C. F. J. Pun, G. Zhou, X. Cheng, and M. L. V. Tse, "Special structured polymer fibers for sensing applications," *Opt. Fiber Technol.*, vol. 16, no. 6, pp. 357–366, 2010.
140. I.-L. Bundalo, R. Lwin, S. Leon-Saval, and A. Argyros, "All-plastic fiber-based pressure sensor," *Appl. Opt.*, vol. 55, no. 4, pp. 811–816, 2016.
141. G. Yan, A. Markov, Y. Chinifooroshan, S. M. Tripathi, W. J. Bock, and M. Skorobogatiy, "Resonant THz sensor for paper quality monitoring using THz fiber Bragg gratings," *Opt. Lett.*, vol. 38, no. 13, pp. 2200–2202, 2013.

4 Sensing with Polymer Fiber Bragg Gratings

4.1 INTRODUCTION

Fiber Bragg gratings (FBGs) are fiber optic structures widely used in sensor applications for the detection of many parameters such as strain, temperature, pressure, humidity, bend, acceleration, vibration, electric and magnetic field, chemical species, among others. FBGs can be considered as discrete sensors but they have the possibility to be multiplexed and offer quasi-distributed sensing. Compared with traditional electric sensors they can provide advantages such as low weight, small size, passive/low power, immunity to electromagnetic interference, the possibility to operate in harsh environments, long distance, etc. Furthermore, FBGs can be embedded in different materials allowing the creation of "smart structures" capable to monitor different properties. Examples of this are the inspection of the curing process during the manufacturing process or also the health monitoring of structures, such as buildings, bridges dams, etc., allowing to get information of the history and status of the structure, preventing maintenance costs and improving safety. Additionally, coating or packaging FBGs in responsive materials can extend their range of sensing capabilities. In view of all these special characteristics, FBGs have been reported in a variety of disciplines, ranging from medicine to civil engineering.

The general principle of operation of an FBG sensor is based on the measurement of the Bragg wavelength shift that occurs when the grating is subjected to an external parameter. The wavelength change could be due to a modification of the physical properties of the grating (length change) or due to an induced effective refractive index change.

Polymer optical fibers (POFs) have attracted much interest among application engineers and scientists due to their unique advantages compared to their silica counterparts as presented before. Among them are their flexibility, high fracture resistance, large strain measurement, low Young's modulus, low densities, non-brittleness, and biocompatibility. Furthermore, if used to monitor compliant structures (e.g., textiles), POFs are much more advantageous than silica fibers. This occurs because the stiff nature of the silica fiber can reinforce the structure. POFs can be fabricated in a wide variety of materials and thus, they can be tailored for the intended application. For instance, POFs can be made of polymer materials with different water absorption capabilities, allowing to produce humidity sensitive and insensitive sensors. Additionally, POFs are composed of organic materials, allowing the fabrication of sensors using different chemical processing techniques.

FBGs in POFs were first demonstrated two decades ago and since then, the technology has been improved along the years. In the same way, as for silica FBGs, polymer optical fiber Bragg gratings (POFBGs) can also cover the same range of applications while taking opportunity of the special properties offered by polymer materials. Therefore, many practical applications have been developed and some of them will be reviewed in this chapter.

FBGs can be used for sensing by monitoring the dependence of the Bragg reflection wavelength with an external parameter, designated here as X, (e.g., strain (ε), temperature (T), hydrostatic pressure (P), humidity (RH), or refractive index of the surrounding medium (n_{cl})). The functional dependence of the Bragg wavelength can be calculated following the Bragg relation shown in Equation 2.1, as [1]:

$$\lambda_{\text{Bragg}} = 2n_{\text{eff}}\Lambda$$

$$\frac{d\lambda_{\text{Bragg}}}{dX} = 2\frac{d}{dX}\left(n_{\text{eff}}\Lambda\right) \tag{4.1}$$

$$= 2\left(\Lambda\frac{dn_{\text{eff}}}{dX} + n_{\text{eff}}\frac{d\Lambda}{dX}\right)$$

which may be expressed as:

$$\Delta\lambda_X = \lambda_{\text{Bragg}}\left(\frac{1}{\Lambda}\frac{d\Lambda}{dX} + \frac{1}{n_{\text{eff}}}\frac{dn_{\text{eff}}}{dX}\right)\Delta X \tag{4.2}$$

where λ_{Bragg} is the initial Bragg wavelength while the first and second term in parentheses are the expansion coefficient and the induced refractive index change due to the external parameter, respectively.

4.2   POFBG STRAIN SENSING

When a longitudinal strain is applied to an FBG, the resonance Bragg wavelength increases following Equation 4.2, as:

$$\Delta\lambda_\varepsilon = \lambda_{\text{Bragg}}\left(\frac{1}{\Lambda}\frac{d\Lambda}{d\varepsilon} + \frac{1}{n_{\text{eff}}}\frac{dn_{\text{eff}}}{d\varepsilon}\right)\Delta\varepsilon \tag{4.3}$$

The first term accounts for the physical length change of the grating period with strain and simply equates to:

$$\frac{1}{\Lambda}\frac{d\Lambda}{d\varepsilon} = 1 \tag{4.4}$$

The second term in Equation 4.3, accounts for the refractive index change due to the strain optic effect. Assuming that the fiber is isotropic, the refractive index change due to the strain optic effect is given by:

$$\frac{1}{n_{\text{eff}}}\frac{dn_{\text{eff}}}{d\varepsilon} = -\frac{n_{\text{eff}}^2}{2}\left(p_{12} - v\left(p_{11} + p_{12}\right)\right) = -p_e \tag{4.5}$$

where p_e is the photoelastic coefficient of the material, p_{11} and p_{12} are the Pockel's (piezo) coefficients of the stress-optic tensor, and ν is the Poisson's ratio. Using Equations 4.4 and 4.5 into Equation 4.3, the Bragg wavelength shift can be rewritten as:

$$\Delta\lambda_\varepsilon = \lambda_{\text{Bragg}}\left(1 - p_e\right)\Delta\varepsilon \qquad (4.6)$$

Thus, the correspondent wavelength-strain sensitivity is expressed from Equation 4.6 as:

$$S_\varepsilon = \lambda_{\text{Bragg}}\left(1 - p_e\right) \qquad (4.7)$$

For a POF composed of polymethylmethacrylate (PMMA), $p_{11} = 0.300$ and $p_{12} = 0.297$ [2], $\nu = 0.35$ to 0.40 [3]. In this way, considering $n_{\text{eff}}(1{,}550 \text{ nm}) = 1.4794$ (obtained from the Sellmeier coefficients given in Ref. [4]), the calculated photoelastic coefficient is 0.0964. Thus, using Equation 4.7, the theoretical strain sensitivity of a PMMA-based POFBG operating at the 1,550 nm region will be equal to 1.4 pm/με. For comparison purposes, in a silica FBG, $p_{11} = 0.113$, $p_{12} = 0.252$, $\nu = 0.16$ and $n_{\text{eff}}(1{,}550\text{nm}) = 1.482$, giving $p_e = 0.213$ and a strain sensitivity value of 1.2 pm/με [5], which is close but smaller than the one theoretically found for a PMMA fiber.

4.2.1 POFBG STRAIN CHARACTERIZATION

The procedure used to characterize an FBG requires its fixation between two mechanical stages (one stationary and another free to move). To hold the fiber, it is commonly used cyanoacrylate or epoxy glue. The FBG is then stretched with pre-defined steps of strain and the Bragg reflection wavelength is acquired for each of the strain steps, using an interrogation system. An example of the setup used for the FBG strain characterization may be seen in Figure 4.1.

Due to the cross sensitive issues to other parameters, such as temperature, pressure, humidity, etc., the environmental conditions should be maintained constant when performing the measurements. The FBG strain sensitivity is acquired when the Bragg wavelength shifts linearly with strain; hence, the strain characterization needs to be performed in the linear elastic region of the material. In this case, POFs have a superior performance than silica optical fibers (ε_{Lim}(POF) >20 mε [6] > ε_{Lim}(Silica) < 6 mε [7]). The typical POFBG spectra for an FBG written in a PMMA microstructured polymer optical fiber (mPOF) as well as the correspondent

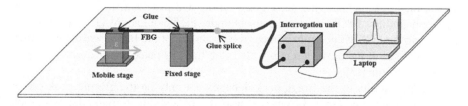

FIGURE 4.1 Schematic used for the strain characterization of a POFBG, where one end of the fiber is fixed and the other is strained with a mechanical linear stage.

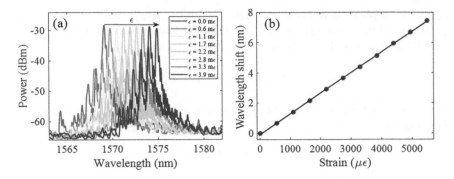

FIGURE 4.2 (a) Spectra acquired for a POFBG written in a PMMA mPOF, for different strain conditions. (b) Corresponding Bragg wavelength shift as a function of applied strain.

peak Bragg wavelength shift acquired for different strain conditions may be seen in Figure 4.2a and b, respectively.

By adjusting a linear regression model to the Bragg peak wavelength shift presented in Figure 4.2b, it is possible to quantify the FBG strain sensitivity.

Since the first inscription of a POFBG, together with its first demonstration as a strain sensor [8], several other works have been reported showing FBG strain sensitivities for fibers of different materials, structures, and spectral regions. A summary of some of those strain sensitivity results may be seen in Table 4.1.

In Table 4.1, it can be seen that most of the strain characterization was performed for the wavelengths between 850 and 1,550 nm. The reason is due to the availability of components at the infrared region. Nevertheless, the most preferred wavelength region that is found around 600 nm (excluding the CYTOP® case), has found little research, being the main reason related with the cost of phase masks and sources for this spectral region. Despite the benefit of lower loss at shorter wavelength regions, this comes with the cost of lower strain sensitivities (i.e., $S_\varepsilon(\sim600$ nm$) \sim 0.6$ pm/µε [9] vs. $S_\varepsilon(\sim1{,}550$ nm$) \sim 1.5$ pm/µε [12]).

In an FBG sensor, it is important that the Bragg peak power remains constant as a function of the external parameter, allowing software tracking methods to easily monitor the Bragg peak wavelength shift. Non-annealed POFBGs show almost no changes on the peak power under low values of strain. However, when strain reaches values above 25 mε the peak power decreases due to the mode propagation losses arising from the elongation of the fiber and reduction of the core diameter [13]. In such elongation regimes, the peak power splits and also vanishes, compromising the applicability of POFBG strain sensors under high elongations [13,23]. However, annealing the POFBG not only reduces the hysteresis of the Bragg wavelength shift, as we will see in the next section but also stabilizes its peak power even for strains as high as 65 mε [13].

The traceability of the Bragg peak depends not only on the grating peak power but also on its bandwidth. In cases where the bandwidth is high, the resolution of the parameter being measured will deteriorate. Because of that, Statkiewicz-Barabach et al. reported the fabrication of a Bragg grating-based Fabry–Pérot (FP)

TABLE 4.1
Summary of POFBG Strain Sensitivities

Fiber	Material	Wavelength (nm)	Annealing	Sensitivity (pm/µε)
SM-mPOF	PMMA	~619	85°C, 2 days	~0.6 [9]
SM-mPOF	PMMA	~847	No	~0.7 [10]
FM-mPOF	PMMA	~962	No	~0.9 [11]
SM-mPOF	PMMA	~1,550	No	~1.3 [10]
SM-SI	PMMA	~1,576	No	~1.5 [8]
SM-SI	PMMA	~1,523	No	~1.5 [12]
SM-SI	PMMA/PS	~854	No	~0.7 [10]
		~1,550		~1.3 [10]
SM-SI	PMMA/PS	~1,550	No	~1.3 [13]
		~1,530	80°C, 2 days	~1.4 [13]
SM-mPOF	PMMA-BDK	~844	60°C, 6 hours	~0.8 [14]
SM-mPOF	PC	~876	~130°C, 36 hours	~0.7 [15]
SM-mPOF	TOPAS® 8007	~877	No	~0.6 [16]
SM-mPOF	TOPAS® 5013	~853	80°C, 8 hours	0.8 [17]
MM-unclad	ZEONEX® 480R	~1,567	100°C, 24 hours	~1.5 [18]
SM-mPOF	ZEONEX® 480R	~829	120°C, 36 hours	~0.8 [19]
MM-unclad	CYTOP®	~604	No	~0.5 [20]
MM-unclad	CYTOP®	~1,435	No	~1.5 [21]
MM (GI-POF)	CYTOP®+PC	~895	No	~0.7 [22]
		~1,249		~1.1 [22]
		~1,561		~1.4 [22]

interferometer allowing to get modulated spectral fringes with bandwidths between 50 pm and 100 pm [24]. With such grating structures, they were able to increase the strain resolution from 203 µε for a standard POFBG to 57 µε for one of the interference fringes of the Bragg grating-based FP cavity.

The bandwidth of common fiber Bragg gratings (either silica or polymer) is intrinsically immune to external perturbations, such as strain and temperature. However, the bandwidth can be made to be strain-dependent if a strain gradient is introduced along the grating length [25]. This strain gradient bandwidth dependence is singular and thus temperature-independent strain sensors are possible. This has been first demonstrated for silica fibers, where a chirped grating was written in a tapered fiber by differential etching, allowing to tailor the grating to become linearly chirped when in tension [25]. Recently, Min et al. [26], motivated by the high elongation opportunities presented by POFs, used the same approach with a chirped FBG in a POF. Results revealed that the central Bragg wavelength was linearly shifted with strain, temperature, and humidity, showing sensitivities similar to the ones reported for standard POFBGs. Furthermore, they also observe that the bandwidth increased with the strain (linear increase from 0.2 nm when unstrained to about 3 nm when under 17.5 mε) while showing temperature insensitiveness as it occurs with chirped silica FBGs. Concerning humidity tests, they revealed a decrease of the bandwidth with increasing humidity (−4.2 pm/%RH) [26]. Regarding this last parameter, one may assume that the best approach to develop a temperature insensitive POFBG sensor may be through the use of POFs made of humidity insensitive materials, such as ZEONEX® or TOPAS®.

4.2.2 POFBGs AND VISCOELASTICITY

The strain sensitivities presented in Table 4.1 for each spectral region are quite similar. However, small discrepancies may be found due to the different fiber materials as well as the thermal history of the fiber. The latter is related with the fiber drawing temperature on the production process [27] and also associated with the post-annealing procedure that is known to act as a chain stress release along the fiber length, allowing the molecules to relax to their random orientation [27,28]. Therefore, annealing treatments of POFBGs have demonstrated better strain responses [28,29]. POFBGs often exhibit hysteresis in the response of Bragg wavelength to strain, particularly when exposed to high levels of strain [13,30,31]. For that reason, fibers have been annealed [13,31], pre-strained [31] and bonded directly to a substrate [31,32]. Regarding the annealing procedures, Yuan et al. showed that the quasi-static linear strain regime can be extended from 28 mε for an unannealed PMMA POF (step-index single-mode (SI-SM), (MORPOF02)) up to values of 38 mε for the same POF annealed up to 80°C during two days [13]. However, results shown later by Abang et al. for a PMMA mPOF [31] revealed hysteretic behaviors for both unannealed and annealed fibers (80°C 48 hours) for strain values of 28 mε. Results revealed that the hysteresis can be effectively reduced by annealing the fiber but cannot be removed [31]. Nevertheless, the different results achieved for each group was, according to Abang et al., due to the differences in the fiber types involved.

The above discussion is intrinsically related to the viscoelastic nature of polymers, which means that the elastic behavior is time-dependent. This phenomenon has been described in literature by different authors [9,11,31–37], and research on this topic has unveiled some restrictions of POFBGs as strain sensors [34,36] as well as potential solutions to solve them [9,31,32].

Viscoelasticity occurs due to the molecular rearrangement of the polymer chains, which dissipates part of the accumulated energy as plastic deformation [38]. The phenomenological description of this time dependence may be visualized through the scheme described in Figure 4.3. On the left-hand side of this figure, it is represented a step of stress being applied to a polymer material and on the right the corresponding strain time response. In this schematic, it is assumed that the amount of stress is below the yield strength. Upon an instantaneous step of strain, the polymer material responds with an immediate strain (ε_1). Then, if the stress duration (T_1) is long, a progressive deformation occurs up to the saturation. This phenomenon arises due to the stress relaxation of the polymer material, leading to a gradual increase in the strain. After the stress being abruptly removed, the polymer material reacts instantaneously (ε_2) and then slowly decays (ε_3) to the initial state

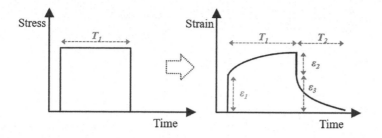

FIGURE 4.3 Representation of the strain time response of a viscoelastic material (right-hand side) upon a step of stress (left-hand side).

after a time period T_2 [38], which has been described to be several hours for large strain values [11,33,35]. Because of this phenomenon, POFBG strain sensors need to be carefully evaluated with respect to their final application since they can present hysteresis on their strain response.

The behavior of POFs under low elongations and low-frequency excitations has already been described in Ref. [34], showing weak viscoelastic behavior. Despite that, their use as accelerometers, microphones, and sensors for fast vibrations detection is not problematic as already explored in POFBGs [34,39,40]. However, it could pose restrictions in situations where an instantaneous read of the strain needs to be accessed after the removal of a load that has been applied to the POF for a long period of time. Studies on this subject revealed that this situation can be partially solved either by bonding the entire fiber to a substrate [31,32] or by pre-straining the fiber [9,31]. Regarding the former, a complete linear behavior without hysteresis was accomplished by bonding the POFBG directly to a plastic cantilever, allowing to have linear responses for strains up to 28 mε [31,32]. Such configuration can only be performed when the sensor is allowed to be embedded or surface bonded to the substrate. In such cases, the fiber is forced to follow the elasticity of the material under analysis, being any residual hysteresis due to the mechanical properties of the substrate.

On what concerns the case where the fiber is pre-strained prior to the sensor operation, Abang et al. [31] showed that in a quasi-static operation and at high strains (i.e., 28 mε), the POFBG will show hysteresis, but in a lower degree when compared with the case without pre-strain. Later on, Bundalo et al. [9] performed a series of strain-relaxation experiments in which the POFBG was cycled ten times for three strain values: 3, 6.5, and 9 mε, being each test implemented for different strain durations: $T_1 = 0.5, 2.5, 5, 10,$ and 50 minutes. Results revealed that any hysteresis disappeared after a certain number of strain-relaxation cycles, where the slow response saturated and the fast response (ε_2) remained always above 65% of the total wavelength interval (see Figure 4.4a). Based on those results, they concluded that if the POFBG is pre-strained to shift the Bragg wavelength sufficiently (more than the upper limit of slow response) then it would always be on the instantaneous ε_2 regime. As a proof of concept, they applied ten strain relaxation cycles between 6 mε (to be sure that the strain was really above the slow response (ε_3)) and 10 mε. The results may be seen in Figure 4.4b, showing that the POFBG can follow instantaneously the applied strain, without any time lag.

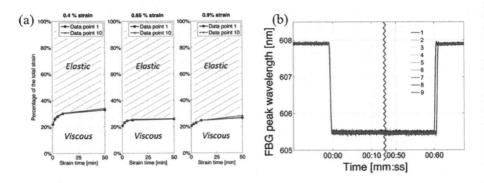

FIGURE 4.4 (a) Percentage of the total strain as function of the strain duration for the strain steps of 4 mε (left), 6.5 mε (center), and 9 mε (right). The curves represent the data points collected for the first and last cycle. (b) Response of the POFBG on the elastic region, done for ten strain relaxation cycles between 6 and 10 mε, showing an instantaneous POFBG response. (Reproduced from I.-L. Bundalo, et al., "Long-term strain response of polymer optical fiber FBG sensors," *Opt. Mater. Express*, vol. 7, no. 3, pp. 967–976, 2017, with permission from the Optical Society of America.)

4.2.3 POFBGs Strain Sensitivity Improvement

According to Equation 4.7, the theoretical strain sensitivity of a PMMA-based POFBG operating at the 1,550 nm region is ~1.4 pm/με, a value that is close to the ones shown in Table 4.1 for POFs made of other materials and considering the same wavelength region. As described before, the fiber thermal history can affect this value, allowing to slightly improve the strain sensitivity [13,28,29]. Nevertheless, other techniques have also been used, such as the fiber tapering [30], as well as the use of in-series silica and polymer fibers [41,42].

When an FBG is under stress, the wavelength shift may be calculated by using the stress-strain relation in Equation 4.6, allowing to obtain:

$$\Delta\lambda_\sigma = \lambda_{\text{Bragg}}\left(1-p_e\right)\frac{\Delta\sigma}{E} \tag{4.8}$$

where E defines Young's modulus. Thus, the stress sensitivity may be defined as:

$$S_\sigma = \frac{\Delta\lambda_\sigma}{\Delta\sigma} = \frac{S_\varepsilon}{E} \tag{4.9}$$

Equation 4.9 may also be expressed as a function of force (F), which translates to:

$$S_F = \frac{\Delta\lambda_F}{\Delta F} = \frac{S_\varepsilon}{E\pi a^2} \tag{4.10}$$

where a defines the fiber radius. From Equations 4.9 and 4.10, it is easy to understand that both stress or force sensitivity could be enhanced if a fiber with a smaller Young's modulus is used. POFs have Young's modulus of ~3 GPa [27] which is much smaller than ~70 GPa, usually found for silica fibers [7]. In this way, stress or force sensitivities would be much higher for an FBG written in a POF than in silica fiber. The group of Peng et al. [43,44] experimentally demonstrated that the stress sensitivity in a POFBG was about 421 pm/MPa, which was twenty eight times higher than the ones reported for silica FBGs. Nevertheless, Equation 4.10 also demonstrates that the stress sensitivity is inversely proportional to the fiber radius. Thus, smaller diameter fibers could be used to improve S_F. The first study regarding the force sensitivity improvement through fiber diameter reduction was performed in 1999 with a silica fiber containing an FBG [45]. The work showed that the tuning force to reach the same amount of wavelength shift could be lowered by over an order of magnitude. Furthermore, studies performed on POFs revealed similar behaviors, showing an enhancement on the force sensitivity [29,46,47].

Bhowmik et al. [30,48] revealed that etching POFs promotes a reduction of Young's modulus and assumed that the Poisson's ratio would also be affected. The explanation given for Young's modulus reduction was presumed to be due to the stress relaxation that softened the material. Their results concerning the strain sensitivity, revealed an improvement from 1.3 to 2.1 pm/µε for a fiber reduction from 180 to 43 µm, respectively. However, later studies reported by other researchers revealed that the strain sensitivity is almost unaffected for etched POFs [29]. In view of these contradictory results and taking into account Equation 4.7 and the absence of literature results regarding the effect of the etching on the photoelastic constant of POFs, one may assume that the strain measurements made in Ref. [30] were performed using the etched and non-etched regions of the POF, leading to higher and lower strain densities in the etched and non-etched regions, respectively, as already explored with silica FBGs [49].

In fact, when an optical fiber is subjected to stress, the strain distribution on a specific fiber portion will depend on its length, Young's modulus and diameter, compared to the rest of the fiber. Thus, if one considers to use a strain sensor based on two similar fibers, with equal diameters and lengths but with different Young's modulus, it is not surprising to find a higher strain density in the fiber with the lower Young's modulus. Based on that, and taking into account that silica fibers are most of the times spliced with a hard UV resin to a silica pigtail fiber (see the Subsection 1.4.2), Oliveira et al. [41,42] decided to implement two fiber portions (an in-series silica and polymer fiber) as a strain sensor head. In this way, the higher density of strain was observed for the POF section (containing an FBG) as a consequence of its lower Young's modulus. The schematic setup, as well as the representation of the fiber sensor, may be seen in Figure 4.5.

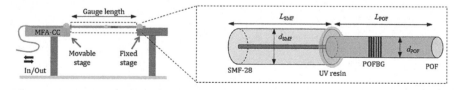

FIGURE 4.5 Schematic strain characterization setup (left-hand side) and the fiber sensing head composed by the silica and polymer fibers (right-hand side). (Reproduced from R. Oliveira, et al., "Strain sensitivity enhancement of a sensing head based on ZEONEX polymer FBG in series with silica fiber," *J. Light. Technol.*, vol. 36, no. 22, pp. 5106–5112, 2018, with permission from IEEE© (2018).)

The mathematical equations used to describe the strain as well as the sensitivities on the POFBG were based on the work developed in Ref. [49], for a silica FBG in-series with a fused taper. Thus, the strain applied to the entire sensing head is given by:

$$\varepsilon_{\text{Tot}} = \frac{L_{\text{POF}} + L_{\text{SMF}} \dfrac{E_{\text{POF}}}{E_{\text{SMF}}} \left(\dfrac{d_{\text{POF}}}{d_{\text{SMF}}}\right)^2}{L_{\text{POF}} + L_{\text{SMF}}} \varepsilon_{\text{POFBG}} \qquad (4.11)$$

where, L_{POF} and L_{SMF} are the length of POF and silica fiber, respectively, and their sum represents the total gauge length. E_{POF}, d_{POF} and E_{SMF}, d_{SMF}, represent Young's modulus and diameter of the POF and silica fiber, respectively. Hence, the POFBG strain sensitivity of the sensing head ($K_{\varepsilon,\text{POFBG}_{\text{SH}}}$), could be written as:

$$K_{\varepsilon,\text{POFBG}_{\text{SH}}} = K_{\varepsilon,\text{POFBG}} \frac{L_{\text{POF}} + L_{\text{SMF}}}{L_{\text{POF}} + L_{\text{SMF}} \dfrac{E_{\text{POF}}}{E_{\text{SMF}}} \left(\dfrac{d_{\text{POF}}}{d_{\text{SMF}}}\right)^2} \qquad (4.12)$$

Here $K_{\varepsilon,\text{POFBG}}$, is the strain sensitivity of the POFBG alone. From this equation, one can realize that because $E_{\text{POF}} \ll E_{\text{SMF}}$, $K_{\varepsilon,\text{POFBG}_{\text{SH}}}$ is improved related to $K_{\varepsilon,\text{POFBG}}$, with a factor of ~$(1 + L_{\text{SMF}}/L_{\text{POF}})$. Nevertheless, if one considers to increase the length of the silica SMF by keeping the same POF length then the strain sensitivity can be further increased. The theoretical results obtained from Equation 4.12 as well as the experimental ones considering an FBG in an unclad ZEONEX® 480R POF ($L_{\text{POF}} = 2.6$ cm, $d_{\text{POF}} = 73.5$ μm and $E_{\text{POF}} = 2.2$ GPa), in combination with a SMF-28 ($d_{\text{SMF}} = 125$ μm, $E_{\text{SMF}} = 70$ GPa), for different gauge lengths (increasing L_{SMF}), may be seen in Figure 4.6.

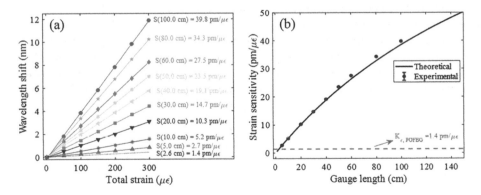

FIGURE 4.6 (a) Experimental results of the Bragg wavelength shift as a function of total strain considering 2.6 cm POF and different gauge lengths (increasing the silica fiber length). (b) Theoretical (from Equation 4.2) and experimental strain sensitivities as function of the gauge length. (Reproduced from R. Oliveira, et al., "Strain sensitivity enhancement of a sensing head based on ZEONEX polymer FBG in series with silica fiber," *J. Light. Technol.*, vol. 36, no. 22, pp. 5106–5112, 2018, with permission from IEEE© (2018).)

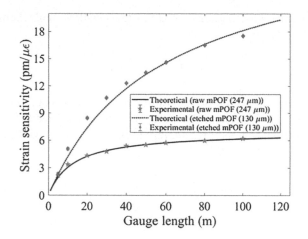

FIGURE 4.7 Experimental and theoretical (calculated from Equation 4.12), strain sensitivity results, for a fiber sensing head composed of 2.7 cm PMMA mPOF in-series with silica fiber, for different gauge lengths (increasing the silica fiber length). The results are shown either for the sensing head composed of the raw mPOF with diameter of 247 μm (star marker points) or etched mPOF with diameter of 130 μm (diamond marker points). (Reprinted with permission from R. Oliveira, et al., "Strain sensitivity control of an in-series silica and polymer FBG," *Sensors*, vol. 18, no. 6, p. 1884, 2018.

As observed from Figure 4.6a and b, $K_{\varepsilon,\text{POFBG}_{\text{SH}}}$, increases with the increasing length of the silica fiber. A figure of merit is observed when $L_{\text{SMF}} \sim L_{\text{POF}}$, where $K_{\varepsilon,\text{POFBG}_{\text{SH}}}$ reaches twice the value of the POFBG alone (i.e., 1.4 pm/με vs. 2.7 pm/ ε). Nevertheless, values as high as 39.8 pm/με were found for a gauge length of 100 cm. Additionally to Young's modulus and fiber lengths, the strain sensitivity defined in Equation 4.12 also depends on the diameter ratio. In this way, reducing the diameter of the POF can further improve the POFBG strain sensitivity results. A detailed work related to this subject may be found in Ref. [42] where the fiber sensing head is composed of a silica fiber in-series with a PMMA mPOF (containing an FBG), either non-etched with diameter of 247 μm or etched to a diameter of 130 μm. A comparison of the theoretical results obtained from Equation 4.12 with the experimental ones may be seen in Figure 4.7 showing that the etching of the POF is an effective method for the increase of the POFBG strain sensitivity.

4.2.4 STRAIN SENSING WITH LPGs IN POFs

When comparing FBGs with long-period gratings (LPGs), one may understand that the much higher period of the latter makes this technology much more attractive. Because of that, easier fabrication technologies such as the mechanical imprinting method or the amplitude mask method could be used in a more affordable way. In view of these fabrication advantages and taking into account the superior mechanical properties of POFs related to silica fibers, a lot of research has been devoted to this type of fiber optic sensor [33,50–55]. LPGs express loss features for wavelengths where the core mode couples to a cladding mode, being the phase-matching condition given by:

$$m\lambda = \left(n_{co} - n_{cl}^i\right)\Lambda_{LPG} \tag{4.13}$$

where Λ_{LPG}, defines the period of the grating, m defines the coupling order and n_{co} and n_{cl} are the effective indices of the core mode and the ith cladding mode. By differentiating Equation 4.13 and rearranging it, one can show how these wavelength regions change with strain [33]:

$$\frac{d\lambda}{d\varepsilon} = \frac{\partial\lambda}{\partial(\Delta n_{eff})}\left(\frac{dn_{co}}{d\varepsilon} - \frac{dn_{cl}^i}{d\varepsilon}\right) + \Lambda_{LPG}\frac{\partial\lambda}{\partial\Lambda_{LPG}} \tag{4.14}$$

where $\Delta n_{eff} = n_{core} - n_{cl}^i$. From Equation 4.14, it can be understood that the contributions to the strain sensitivity are either due to the change in the differential effective index (Δn_{eff}) and also due to the change in the grating periodicity (Λ_{LPG}), usually referred to as the material and waveguide effects, respectively. The material contribution results from the strain-optic effect (change in refractive index) and Poisson's effect (change in transverse dimensions), while the waveguide contribution depends on the slope of the characteristic curve of the resonance band, $\partial\lambda/\partial\Lambda_{LPG}$ [56]. The waveguide effect is normally the most pronounced effect in determining the strain response of the grating in polymer fibers. Furthermore, the overall shift in a resonance band due to a particular magnitude of axial strain is a function of the grating period and the order of the corresponding cladding mode [56]. Strain sensitivities of LPGs written under different technologies, polymer materials, fiber structures, periods and with resonant dip wavelengths between 500 and 900 nm, have already been shown in literature and a summary of some of them may be seen in Table 4.2.

TABLE 4.2
Compilation of LPG's Strain Sensitivities in POFs

Fiber	Material	Fabrication Technology	Dip Location (nm)	Sensitivity (pm/µε)
SM-mPOF	PMMA	Temperature molded	~570	−11.8 [33]
SM-mPOF	PMMA + PC	Temperature molded	~580	−11.0 [55]
SM-mPOF	PMMA	Temperature molded	~595	−0.5 [50]
SM-mPOF	PMMA	Temperature molded	~684	−1.4 [52]
SM-mPOF	PMMA doped w/BDK	UV (325 nm)	~777	−1.6 [57]
SM-mPOF	PMMA + PC	CO_2	~664	−10.0 [58]
SM-mPOF	PMMA doped w/BDK	UV (248 nm)	~870	−2.3 [59]

As can be observed from Table 4.2, the LPG strain sensitivity values are negative and present values ranging between 0.5 and 11.8 pm/με. The discrepancies may be attributed to the fiber design, fiber material, thermal history of the fiber, periodicity of the fiber and order of the corresponding cladding mode.

Regarding the applications, LPGs have been explored on the assessment of respiratory movements by embedding them in an elastic textile fabric [51]. Preliminary tests regarding the use of LPGs for the detection and measurement of the strain rate and magnitude of engineering structures has been described in Ref. [50] by surface bonding an LPG to a steel plate. Also, the fabrication of a pressure sensor intended for future endoscopic measurements has been reported in Ref. [58], where the LPG is attached to a pod-like structure that converts pressure into longitudinal strain. By doing that, it was possible to achieve a pressure sensitivity of 105 nm/MPa in a range of 0–15 kPa [58].

4.3   POFBG TEMPERATURE SENSING

Temperature is a physical quantity used to provide valuable information in a variety of applications, which include climate research, food technology, among others. The capability to track this parameter can help, for instance, to save energy and monitor the production/efficiency of different processes. Conventional temperature detection schemes include heat resistance, infrared radiation, thermocouple, etc. However, such sensors are influenced by electromagnetic interference or cannot be used in harsh environments. Fiber optic sensors can solve some of these problems and therefore be considered as a viable option over conventional sensors. FBGs are well-known from their wavelength division multiplexing capabilities and thus, they can be used as single or quasi-distributed temperature sensors. FBG temperature sensors have been mainly described using silica fibers which is now a mature technology. Nevertheless, POFBGs are also attractive due to their negative and higher thermo-optic coefficient, which impose a negative and better temperature sensitivity than the one achieved for the conventional FBG fiber.

After the first demonstration of the inscription of POFBGs as well as the first demonstration of their applications as strain sensors, in 1999, by Peng et al. [8,60], the same group was also pioneered in demonstrating the first temperature characterization of a POFBG in 2001 [61]. The characterization was performed in a PMMA SM-POF via Peltier element with temperatures ranging between 20°C and 80°C. A nonlinear behavior was obtained with a rough temperature sensitivity of −360 pm/°C, presenting no hysteresis in both increasing and decreasing temperature. A subsequent work made by the same group and with the same fiber reported a linear behavior with a sensitivity of −149 pm/°C and without hysteresis for temperatures ranging between 20°C and 65°C [12]. These findings compare very favorably with the typical value of ~10 pm/°C reported for Bragg gratings written in silica fibers. Conversely, in the subsequent years, further studies on the same subject made by other groups revealed that the mechanism behind the temperature sensing is far more complex. This occurs because the control in which the humidity tests are being performed [62–64] as well as the thermal treatments before the characterization [65] are essential for the performance of the POFBG temperature sensor.

4.3.1 Theoretical POFBG Temperature Sensitivity

When an FBG is exposed to a temperature variation (ΔT) at constant strain and humidity conditions, the Bragg wavelength will shift according to Equation 4.2, as:

$$\Delta \lambda_T = \lambda_{Bragg}\left(\frac{1}{\Lambda}\frac{d\Lambda}{dT} + \frac{1}{n_{eff}}\frac{dn_{eff}}{dT}\right)\Delta T \qquad (4.15)$$

where $(1/\Lambda)(d\Lambda/dT)$ is the normalized thermal-expansion coefficient defined as α, while $(1/n_{eff})(dn_{eff}/dT)$ is the normalized thermo-optic coefficient, normally referred to as ξ. The theoretical wavelength-temperature sensitivity of an FBG can thus be expressed in a short form as:

$$S_T = \lambda_{Bragg}(\alpha + \xi) \qquad (4.16)$$

Thus, the thermal sensitivity of an FBG is dependent on the sign and magnitude of α and ξ. Generally, polymer materials have $\alpha > 0$, but $\xi < 0$, contrary to silica. Thus, while the sensitivity of a silica FBG is the result of the contribution of both positive coefficients, the same does not apply for POFBGs. Instead, the higher the difference between the two coefficients, the higher the temperature sensitivity. A comparison between the magnitude of α and ξ, for different transparent polymer optical materials, is shown in Figure 4.8. From this figure, one can see that the magnitude of α is higher than that of ξ only for perfluorinated (PF) material and thus, an FBG written in a POF made of this material would provide a positive temperature sensitivity, as reported elsewhere [20,21]. However, this is one particular case, since most of the polymer materials present $\xi > \alpha$.

From Equation 4.16 and taking as an example the thermal coefficients of the mostly used POF material (PMMA), $\alpha = 7.2 \times 10^{-5}/°C$ and $\xi = -8.5 \times 10^{-5}/°C$ [66] and considering an FBG written at the 1,550 nm region, the theoretical sensitivity would be -20.2 pm/°C. This value is approximately 1.4 times higher than the one theoretically

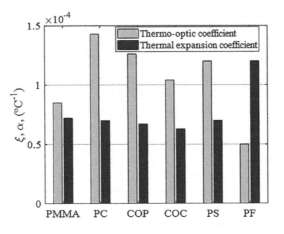

FIGURE 4.8 Comparison between the magnitude of the thermo-optic (ξ) and thermal expansion (α) coefficients of different transparent POF materials, obtained from Refs. [66,67].

estimated for a germanium doped silica fiber Bragg grating which is approximately 14.2 pm/°C ($\alpha = 0.55 \times 10^{-6}/°C$ and $\xi = 8.6 \times 10^{-6}/°C$) [5]. Nevertheless, looking to Figure 4.8, one can see that the thermal sensitivity can be further increased if polymers such as polycarbonate (PC) or cycloolefin polymers (COP) are chosen as the host materials of POFs. In fact, there exist literature reports where POFBG temperature sensitivities can reach values as high as −520 pm/°C at 1,550 nm for a POF made of bisphenol-A acrylate [68,69]. Results of this magnitude may only be understood when $\xi \gg \alpha$.

Temperature characterizations of POFBGs have been accomplished in a variety of POF materials, fiber types, and wavelength regions. Examples of some of the sensitivities reached in literature may be found in Table 4.3.

TABLE 4.3
Summary of POFBG Temperature Sensitivities for Different POFs at Different Wavelengths

Fiber	Material	Wavelength (nm)	Annealing	Humidity	Sensitivity (pm/°C)
SM-SI	PMMA	~1,550	No	Ambient	−149 [12]
SM-mPOF	PMMA	~1,550	65°C, 24 hours (dry)	Ambient	~−71 [70]
SM-SI	Copolymer of styrene and MMA	~1,550	No	Dry	~−10
				Water	~−35
				Ambient	~−138 [64]
SM-mPOF	BDK doped PMMA	~850	60°C, 15 minutes (water)	Ambient	~−57 [14]
SM SI	BDK doped PMMA	~950	80°C, 48 hours	Ambient	~−58
				60%RH	~−4.3
SM-SI	BDK doped PMMA	~1,550	No	Ambient	~−149 [71]
MM-SI	bisphenol-A acrylate	~1,550	60°C, 12 hours (dry)	48%RH	~−520 [68,69]
SM-mPOF	PC	~850	125°C, 36 hours (dry)	90%RH	~−26 [72]
SM-mPOF	TOPAS® 8007	~1,550	No	55%RH	~−37 [73]
SM-mPOF	TOPAS® 8007	~850	No	Ambient	~−78 [16]
SM-mPOF	TOPAS® 5013	~850	80°C, 3 hours (dry)	Ambient	~−15 [17]
MM-unclad	ZEONEX® 480R	~1,550	100°C, 24 hours (dry)	80%RH	~−65 [18]
SM-mPOF	ZEONEX® 480R	~850	120°C, 36 hours (dry)	50%RH	~−24 [19]
SM-mPOF	ZEONEX® & PMMA	~850	85°C, 24 hours, (90%RH)	50%RH	~−15 [74]
MM-unclad	CYTOP®	~1,550	No	50%RH	~+28 [21]
MM (GI-POF)	CYTOP® & PC	~1,550	No	48%RH	~+26 [75]
MM-unclad	CYTOP®	~600	No	Ambient	~+11 [20]

Despite the values found for bulk polymer materials being relatively well-known, the reported POFBG temperature sensitivities have been described in a wide range of values, even among POFs of the same material and in the same wavelength region [63–65]. The complexity of such behavior has been studied in the last few years and it has been shown that depends on the thermal history of the fiber as well as on the humidity conditions in which the sensor operates. Those characteristics will be addressed in the following paragraphs.

4.3.2 POFBG CHARACTERIZATION AND SENSITIVITIES

In general, the sensitivity to temperature of POFBGs can be made either by placing the POFBG on the vicinity of a thermoelectric cooler (TEC) system or by placing the fiber inside a climatic chamber. Examples of these two systems may be seen in Figure 4.9a and b, respectively.

The simplicity and low-cost of a TEC device led most of the research community to use this device for the temperature characterizations. The process is made with increasing and/or decreasing temperature steps. For each temperature step, there is an idle period to allow the stabilization of the Bragg peak wavelength. The typical Bragg reflection spectrum as well as the peak wavelength as function of temperature for a non-annealed PMMA mPOF characterized in a TEC device at room humidity conditions may be seen in Figure 4.10a and b, respectively.

From Figure 4.10 it can be seen a blue-shift of the Bragg peak wavelength with increasing temperature. The result was expected since $\xi > \alpha$. Nevertheless, the reflection spectra presented in Figure 4.10a show also the existence of a higher wavelength shift at higher temperatures. This result can be clearly seen when the Bragg wavelength peak is plotted as a function of the temperature, which shows a nonlinear behavior (see Figure 4.10b). This type of performance was reported on the first POFBG temperature characterization; [61] yet, the mechanism behind the effect was not explained. This has only been accomplished five years later through a careful study performed by Carroll and co-workers [65]. For that, a 1,550 nm POFBG written in a flat sides PMMA mPOF was subjected to three temperature cycles. The first cycle went up to 77°C, the second up to 86°C, and the third up to 92°C. After

(a)   (b)

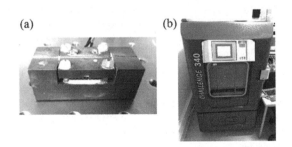

FIGURE 4.9 Conventional systems used for the POFBG temperature characterizations: (a) TEC, capable to provide specific values of temperature, and (b) climatic chamber, capable to simultaneously provide specific values of temperature and humidity.

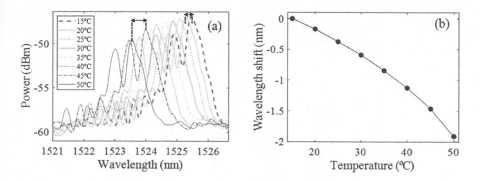

FIGURE 4.10 (a) Bragg reflection spectra and (b) Bragg peak wavelength, as a function of temperature, for a non-annealed PMMA mPOF, characterized in a TEC element in an open laboratory, without humidity control.

reaching the maximum temperature in each cycle, the fiber was allowed to return to room temperature. Results from the first cycle revealed that the Bragg wavelength shift followed a linear behavior for temperatures up to ~50°C, while a fast decrease was seen for temperatures above this value. After returning to room temperature, they noticed that the Bragg wavelength shift was permanently blue-shifted ~8 nm. For the second temperature cycle, the linear behavior had been extended up to the last temperature reached in the first cycle (~77°C), where a fast drop of the Bragg wavelength was then followed. After this cycle, a permanent wavelength shift was observed again (~10 nm). The third test was then performed and the same conclusions reached for the second test were also observed (extension of the linear region up to the value of the highest temperature reached in the prior cycle, followed by a fast drop of the Bragg wavelength). As conclusions, the authors argue that the drop of the Bragg wavelength was the result of the permanent shrinkage of the fiber, creating a permanent blue wavelength shift as a result of the partial release of stress built-in during the heat-drawing process [65]. Nevertheless, a consequence of this stress release was seen as a decrease of the POFBG sensitivity on the linear region of each temperature cycle. To further extend their work, they characterize a POFBG in a fiber that had been pre/annealed at 80°C for 7 hours. The results showed that the operational temperature of the fiber (linear and reversible behavior) could be extended up to 89°C. However, such improvement came with the cost of reduced temperature sensitivity (–88 pm/°C for a non-annealed fiber versus –52 pm/°C for a pre-annealed fiber). Similar conclusions were also observed for SI-PMMA POFBGs at 1,550 nm region by Yuan et al. [13] in a later study of the annealing effects on the strain and temperature performance of POFBGs.

As discussed before, annealing procedures can extend the operating temperature of PMMA-based POFBGs allowing them to reach values up to 89°C [65]. Nonetheless, the value can still be improved if POFs are made of high thermal resistance material. Up to now, there has been some research on the development of POFBGs for high temperatures. Examples of that may be found for POFs made of PC [15], ZEONEX® 480R [41,74] and TOPAS® 5013 [17]. For comparison purposes,

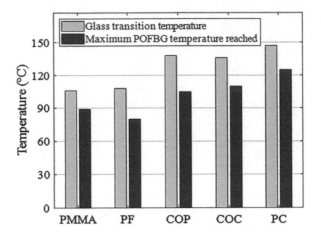

FIGURE 4.11 Glass transition temperature obtained from [66,76] and maximum temperature reached using POFBGs written in different POF-based materials: PMMA, SM-mPOF [65]; PF, MM-unclad [20]; ZEONEX® 480R, MM-unclad [41]; TOPAS® 5013, SM-mPOF [17]; PC, SM-mPOF [15].

the glass transition temperature, as well as the highest POFBG temperature for different POF-based materials is shown in Figure 4.11.

From Figure 4.11 one can observe that the highest POFBG operational temperature is attained for PC (up to 125°C [15]). While these values cannot compete with the ones reported for silica fibers, that are around 500°C for standard FBGs and 1,100°C for regenerated ones, they can still find applications such as the ones found in the automobile industry, specifically at the engine area where temperatures can reach 100°C.

From the previous discussion related to the POFBG thermal annealing, one may assume that fibers made of the same host materials and with similar annealing procedures would present similar thermal sensitivities. However, this is not always the case and the reason is essentially associated with the humidity cross-sensitivity as demonstrated by Harbach [62,64] and later by Zhang et al. [63]. This discussion can only be applied for POFs made of materials with affinity to water, as is the case of PMMA, PC, polystyrene (PS), etc. The reason relates with the humidity sensitivity of the polymer optical fiber material that causes absorption/desorption of water, leading to a change on its mechanical and optical properties and inherently on the resonance Bragg wavelength. Considering the case where the temperature surrounding the POFBG is kept constant while an environment step humidity increase is implemented, there exists water absorption from the polymer material causing swelling of the POF which results in: (i) longitudinal expansion of the grating planes and (ii) increase of the refractive index due to the changes in the mean polarizability [62]. Both effects result in a red wavelength shift of the resonance Bragg wavelength. Nevertheless, the effect is reversible upon a step humidity decrease on the same amount. Since most of the POFBG temperature characterizations have been made using Peltier devices in an open laboratory, the humidity surrounding the POF is also involved in the Bragg wavelength shift. However, this was not realized in many of the works published in literature. Thus, the inconsistency of temperature sensitivities was not only due to the thermal history of the fiber, as described previously, but also due to the uncontrolled environment humidity conditions. Furthermore, it is known that the highest POFBG temperature sensitivities were reached in TEC devices [12,61,77] and the reason can be associated with the cumulative effect of the higher thermo-optic coefficient compared to the thermal expansion coefficient, together with the decrease of the moisture vapor pressure surrounding the fiber, which promotes an additional blue-shift of the Bragg wavelength due to the moisture desorption from the POF [63].

From the above discussions, one can understand that the applicability of POFBGs at high temperatures (>50°C), requires a prior annealing procedure at temperatures even higher, allowing to operate the sensor linearly (without hysteresis). Furthermore, regarding POFs based on materials with affinity to water, it is necessary to perform the characterization under constant humidity conditions, allowing to remove the cross-sensitivity issues. From these two conditions, the POFBG temperature sensitivity can be tuned in two ways: either by annealing or not the POF, which allows to obtain low and high sensitivities, respectively (keeping in mind that without annealing the temperature range is limited); or through the use of POFBGs in humidity sensitive materials allowing to have the cumulative effect on the Bragg wavelength shift as a consequence of both temperature and humidity conditions.

Additionally, the temperature sensitivity can also be controlled by other methods. One of them is by pre-straining the POFBG in order to avoid the thermal expansion of the fiber material. Consequently, the POFBG response would only depend on the contribution of the thermo-optic effect [78], allowing to have higher POFBG temperature sensitivity. Another methodology that can be used to improve the thermal sensitivity of POFBG is by etching the POF as demonstrated by Bhowmik et al. for a POFBG inscribed in a PMMA SI SM-POF with core doped with benzyl methacrylate (BzMA) [30,79]. In one of those works, the sensitivity was enhanced from −95 to −170 pm/°C by reducing the original fiber diameter from 180 to 55 μm, respectively. The sensitivity enhancement was described as the result of the reduction of Young's modulus after etching, which increased the value of α [30]. Despite the interesting result, it is worth to mention that the characterization was made in a Peltier device in an open laboratory. Thus, due to the hygroscopic nature of the material that composes the POF, volumetric changes related to the water absorption/desorption should also be taken into account on the observed sensitivities.

The higher tuning capabilities presented by POFBGs when subjected to temperature led Kalli et al. to consider the deposition of thin-film resistive heater on the surface of a POF containing an FBG [80]. The work was done by coating the fiber with a metallic thin film layer of Palladium Copper. The experiment was done with electric current flowing through the thin film metal coating, resulting in Joule heating and increasing the temperature of the fiber and grating. The device had a wavelength shift of ~2 nm for an input power of 160 mW, leading to a wavelength to input power coefficient of −13.4 pm/mW and a time constant of 1.7 s^{-1} [80].

4.4 POFBG HUMIDITY SENSING

Humidity sensors can find numerous applications in climate research, chemical industry, aviation, semiconductors, and food technology. Nowadays, a variety of humidity sensors have been reported, which include the mechanical hygrometer, chilled mirror hygrometer, wet and dry bulb psychrometer, infrared optical absorption hygrometer, and electronic sensors [81]. Yet, technologies based on optical fibers have been proposed due to their special characteristics. However, silica optical fibers are not hygroscopic and thus, they cannot be influenced by humidity. For that reason, materials with hygroscopic properties such as polyimide [82,83], nafion [84], chitosan [85], polyvinyl alcohol [86–89], polyethylene oxide [90], agar [91,92], and photopolymerizable adhesives [93,94] have been incorporated in silica optical fibers through different technologies, such as fiber Bragg gratings [82,83,89,91,92,95], long-period gratings [86,88,95], and fiber-based FP cavities [84,85,87,93,94]. However, the deposition of such materials is sometimes difficult to accomplish. Yet, the availability of POFs with hygroscopic properties makes their use much more attractive for the development of humidity sensors.

It is known that some polymer materials present high hygroscopic properties while others show almost negligible water absorption. A compilation of the amount of water absorption allowed for different polymer materials may be seen in Figure 4.12.

The correct choice of the POF material dictates its final application. Looking at Figure 4.12, materials, such as CYTOP®, TOPAS® 5013 or ZEONEX®480R, have almost negligible water absorption. Thus, their mechanical and optical properties remain almost stable in humidity conditions and thus, they are not good candidates for a POF-based humidity sensor. On the other hand, PMMA shows the highest water absorption capabilities amongst the polymers shown in Figure 4.12 making this material well attractive for this application.

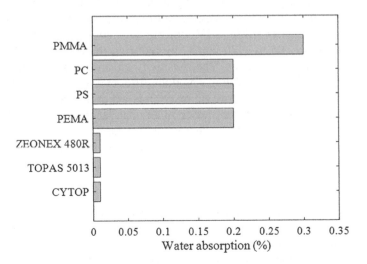

FIGURE 4.12 Water absorption (24 hour/equilibrium) of different transparent polymer optical fiber materials [76,96–101].

A material showing hygroscopic properties will present both volumetric and refractive index changes upon water absorption/desorption. Regarding the former, it can be associated with the swelling process due to the entrance of water into the polymer matrix. The latter can be related to the changes in the mean polarizability of the material. Additionally, the water absorption by hygroscopic polymers leads to the appearance of attenuation bands due to the O-H molecular vibration overtones reducing its transparency in the useful low loss windows. In fact, at 90% relative humidity and temperature of 50°C, the attenuation in some of the low loss regions of PMMA can be increased to values as high as 250 dB/km [102]. These properties can thus be used to develop polymer optical fibers capable to measure humidity using intensity [103], interferometric [104] and wavelength-based techniques [105].

Regarding wavelength-based techniques, which are the most preferred due to the low complexity and reliability involved, Harbach et al. pioneered in works reporting PMMA-based POFBGs as humidity-based sensors [64], [101]. Subsequently, a number of publications regarding this property were published by other researchers using fibers of different geometries, shapes, dimensions, and materials [70,72,74,106–108].

4.4.1 THEORY BEHIND POFBG HUMIDITY SENSORS

When a POFBG made of a material capable to absorb water is in the presence of humidity at a constant temperature, the theoretical sensitivity may be expressed by Equation 4.2 [109], as:

$$S_H = \lambda_{\text{Bragg}} (\beta + \eta) \tag{4.17}$$

where the first term in parentheses represents the swelling coefficient due to the humidity induced volumetric change and the second term denotes the normalized refractive index change due to humidity.

The humidity dependence of the refractive index of the polymer absorbed moisture can be expressed as [110]:

$$\frac{dn_{\text{eff}}}{dH} = \frac{\left(n_{\text{eff}}^2 + 2\right)^2}{6n_{\text{eff}}} k_m S \left(1 - \frac{f}{f_c}\right) \tag{4.18}$$

where k_m is the water molar refraction divided by its molecular weight and S is the moisture solubility of the polymer. The parameter f defines the fraction of absorbed moisture that contributes to the increase in polymer volume, which depends on temperature ($f = 0$ means no change in the polymer volume as it absorbs moisture; $f = 1$ means that the increase of the polymer volume is in the same amount of the absorbed moisture). Finally, f_c is the critical value which is expressed as:

$$f_c = k_m \, \rho_m \frac{n_p^2 + 2}{n_p^2 - 1} \tag{4.19}$$

where ρ_m defines the density of water and n_p the refractive index of the polymer without moisture. Considering PMMA as the host material and using $k_m = 0.2012$ cm³/g,

$\rho_m = 0.9970$ g/cm^3 at 25°C [110], n_{PMMA} (1,550 nm) = 1.4794 at 25°C [4] and using Equation 4.19, one can estimate f_c to be equal to 0.7. Therefore, it is easy to understand that the humidity-refractive index dependence will be positive and negative when $f < 0.7$ and $f > 0.7$, respectively. Considering the values reported for PMMA at 25°C as $f = 0.6$ and $S = 2.54 \times 10^{-4}$ g/cm^3/%RH [110], the humidity-refractive index dependence (η) can be estimated to 1.44×10^{-5} /%RH.

The volumetric change induced by water uptake in a PMMA fiber (β) can be calculated using the swelling coefficient along the fiber direction expressed in [109], as:

$$\beta = \frac{\Delta V}{3V_0} = \frac{\rho_m f w}{300} \tag{4.20}$$

where w is the water absorbed by the polymer fiber material, which for PMMA ranges between 1.6 and 2% of total weight [109]. Using the previously reported parameters, it is possible to calculate the theoretical value of β to be equal to 3.99×10^{-5}/%RH. Since this value is higher than the one presented for the humidity-refractive index dependence ($\eta = 1.44 \times 10^{-5}$/%RH), then, it is the principal contribution for the sensitivity to the relative humidity. Based on the above calculations, the contribution of both volumetric and refractive index change will produce, according to Equation 4.17, a theoretical PMMA-POFBG sensitivity of 68.7 pm/%RH (at 25°C and 1,550 nm).

4.4.2 Experimental Characterization of POFBG Humidity Sensors

The first work reporting humidity sensitivity with a POFBG was made in 2008 using a SI-POF composed of a PMMA cladding and a copolymer of PMMA and PS in the core [62]. Subsequent works performed by other authors revealed the possibility to develop these type of sensors in a variety of polymer optical fiber materials and fiber structures [14,16,18–21,69,70,72,105,111].

As it is known, the refractive index and physical length change of a POFBGs are temperature sensitive and thus, cross-sensitivity issues need to be taken into account [112–114]. In this way, humidity characterizations require that the temperature is constant during the experiment. Whenever a POFBG based on a humidity sensitive polymer material is submitted to a humidity step increase, a positive wavelength shift of the Bragg wavelength is observed. The typical spectral behavior of a PMMA-POFBG, as well as the Bragg wavelength shift attained by different PMMA-based POFs can be seen in Figure 4.13.

From Figure 4.13a, a reduction in the peak power at high humidity levels can be observed, which is caused by the polymer water absorption [195]. Furthermore, it is observed a red wavelength shift of the Bragg wavelength peak, that is attributed to the net effect of the hygral expansion and refractive index change (both positive). As said before, the total Bragg wavelength shift depends on many parameters, namely the fiber material (PMMA, PC, PS, etc.), fiber structure (mPOF, SI, graded-index (GI), etc.), fiber drawing conditions (high/low temperature; high low tension), fiber annealing, operation wavelength, etc. Based on that, different sensitivities have been reported in literature. A summary of some of them may be seen in Table 4.4.

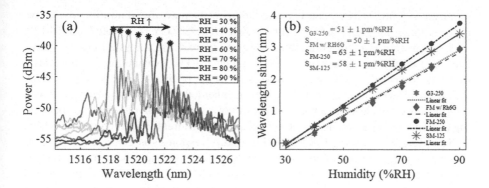

FIGURE 4.13 Bragg wavelength spectra of a PMMA-based mPOF (a) and peak Bragg wavelength shift of different PMMA-based mPOFs (b), at different relative humidity conditions. (Reprinted with permission from R. Nogueira, et al., "New advances in polymer fiber Bragg gratings," *Opt. Laser Technol.*, vol. 78, pp. 104–109, 2016.)

The water absorption capabilities presented by some polymer materials can be a drawback for sensing applications targeting parameters like strain, temperature, pressure, etc. The reason relates to the cross sensitivity to humidity. Despite the possibility of using different fiber optic technologies for the discrimination between

TABLE 4.4
Summary of POFBG Humidity Sensitivity

Fiber Type	Material	Wavelength (nm)	Temperature (°C)	Annealing	Sensitivity (pm/%RH)
SM-mPOF	PMMA	~650	25, 50, 75	80°C; 90%RH (1 day)	~35 [105]
SM-mPOF	PMMA	~1,550	-	65°C (1 day)	~58 [70]
SM-mPOF	BDK doped PMMA	~850	25	60°C in H_2O (15 minutes)	~20 [14]
SM-SI	DPDS doped PMMA	~1,550	40	80°C (2 days)	~54 [115]
SM-SI	TSB doped PMMA	~1,300	20	80°C (2 days)	~66 [63]
SM-mPOF	PC	~850	25, 50, 100	125°C (36 hours) + (100°C; 90%RH (6 hours))	~7 [72]
MM-SI	bisphenol-A acrylate	~1,550	30	60°C (12 hours)	~67 [69]
SM-mPOF	TOPAS® 8007	~850; ~1,550	25	-	~0 [16]
SM-mPOF	ZEONEX® 480R	~850	50	120°C (36 hours)	~0 [19]
MM-unclad	ZEONEX® 480R	~1,550	25	90°C (1 day)	~0 [18]
MM-unclad	CYTOP®	~650	30	-	~0 [20]

parameters [106,111,116], conservative approaches may use humidity insensitive polymer materials [16,117], (ex. TOPAS®, ZEONEX®, or CYTOP® (see Figure 4.12)).

The sensitivities reported in Table 4.4 and Figure 4.13b shows a range of values even for POFBGs composed of the same materials. The explanation for such variation was described by Zhang and co-workers as the result of the anisotropy created during the drawing process that still exists in annealed fibers (i.e., annealing just with temperature). This leads to a wide variety of polymer molecular alignments, promoting different volume expansions and consequently different POFBG responses [109].

The hysteresis observed in POFBG humidity responses is also a concern. This is due to changes in the microscopic structure of the polymer network since short polymer chains need to be broken and other chains have to find their optimal arrangement. Again, the molecular chain alignment during the drawing of the fiber plays an important role in this effect. Traditional fiber annealing treatments use high temperatures without control of the humidity (i.e. considered as low humidity). As described by Zhang et al. [109], chain molecular alignment inside the fiber still exists after this type of procedure and thus, unpredictable humidity sensitivity responses could be achieved. However, if the fiber annealing is performed at high temperature and humidity values, this issue can be solved as described in detail by Woyessa et al. [105]. Nonetheless, similar results were also achieved by Bonefacino et al. [118] by performing ten humidity cycles of 17 hours each, at a temperature of 20°C. In both studies, the hysteresis was solved at a cost of sensitivity reduction.

Another important feature when dealing with POFBGs as humidity sensors relates to the response time, associated to the time needed for the water uptake by the polymer matrix. This property has been associated either with the anisotropy of the fiber left during the heat-drawing process, fiber diameters [107,119], and fiber geometries [70,108,111]. Examples of the Bragg wavelength response under different humidity values, for a PMMA mPOF with 125 μm diameter, as well as a comparison of it with other PMMA mPOFs, for the same humidity step level, may be seen in Figure 4.14a and b, respectively.

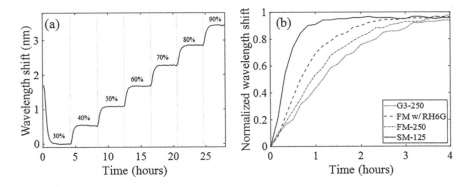

FIGURE 4.14 (a) Time evolution of the Bragg peak wavelength for a 125 μm diameter PMMA-based mPOF, under different step humidity conditions. (b) Time response of POFBGs in PMMA mPOFs of different diameters (G3–250 μm; FM w/ Rh6G-170 μm; FM-250 μm; SM-125 μm) and shapes, under a humidity step rise of 10%RH.

The typical time needed for a POFBG to reach stabilization is in the order of tens of minutes (see Figure 4.14a and b) and depends on the fiber diameter (see Figure 4.14b). Nevertheless, the drawing conditions (which are known to induce different fiber molecular arrangements) have also influenced the fiber response time [107].

Assuming a hygroscopic polymer optical fiber as a homogenous cylindrical rod, the process of water absorption or desorption can be described by the diffusion theory of mass transfer as:

$$\frac{\partial C}{\partial t} = D\frac{\partial^2 C}{dr^2} + \frac{D}{r}\frac{\partial C}{\partial r} \tag{4.21}$$

where D is the diffusion coefficient and $C(t, r)$ is the concentration depending on time (t) and radial position (r). Solving Equation 4.21 with suitable boundary conditions, the normalized concentration for in-diffusion (absorption) can be obtained as follows [120]:

$$C(t,r) = 1 - \frac{2}{a}\sum_{n=1}^{\infty}\frac{\exp\left(-D\alpha_n^2 t\right)J_0\left(r\alpha_n\right)}{\alpha_n J_1\left(a\alpha_n\right)} \tag{4.22}$$

and for out-diffusion (desorption):

$$C(t,r) = \frac{2}{a}\sum_{n=1}^{\infty}\frac{\exp\left(-D\alpha_n^2 t\right)J_0\left(r\alpha_n\right)}{\alpha_n J_1\left(a\alpha_n\right)} \tag{4.23}$$

being a the cylinder radius and α_n the nth positive root of $J_0(a\alpha_n) = 0$. Taking the example of PMMA, the diffusion coefficients for water sorption and desorption are 6.4×10^{-9} cm^2/s and 9×10^{-9} cm^2/s [121], respectively. Therefore, from Equations 4.22 and 4.23, one may see that the only way to improve the response time is through the reduction of the radius of the fiber. For that reason, this has been investigated with different approaches: etching the POF in the FBG region [107,119]; laser micromachining the region containing the FBG with 248 nm UV laser (D-shaped and slotted structures) [108]; and by drawing POFs with flat sides [111]. Regarding the work reported in Ref. [119], which uses an etched microfiber with only 25 µm, it can be found an impressive value of just 4.5 seconds for POFBG signal stabilization upon a humidity step increase. This value contrasts with the tens of minutes described for most of the POFBGs found in the literature. Furthermore, it has been already shown that strained POFBGs can also be used to improve the response time to humidity [108,109]. The response time reported in Ref. [109] was decreased from 25 minutes to just 13 minutes for a POFBG with 200 µm, without and with strain, respectively. In these works, the hygral expansion is not involved for the Bragg wavelength shift and thus, the only contribution is related with the refractive index change, which the authors assumed to be lower than the volumetric change.

Similar to POFBGs, LPGs written in POFs can also be used to act as humidity sensors [53,54,103,122]. The description for the wavelength shift in these fiber structures should follow similar deductions as for POFBGs, showing a red wavelength shift as the result of the net contribution of the volume expansion (separation of the grating planes) and refractive index change. However, this tendency has only been observed for LPGs written by the permanent mechanical imprinting method [122], while a blue wavelength shift has been described for LPGs written by the UV photoinscription process [103]. These differences could be however related to the mode coupled in each method. Regarding the sensitivities, it has been demonstrated values of the order of ~60 pm/%RH [103] for a resonance wavelength located at 680 nm, which is close to the ones reported for POFBGs [69,70].

4.5 POFBG REFRACTIVE INDEX SENSING

The capability to characterize the optical properties of fluids is crucial in several processes, namely salinity of water, fuel quality, biotechnology processes, etc. Standard devices such as the Abbe refractometer are frequently used to measure the refractive index. However, they have limitations regarding their weight and size. Fiber optic sensors are an alternative since they are compact in size and have the ability to be used in quasi-distributed or tip-based sensing applications. Over the last decades, several fiber optic technologies based on silica fibers have been proposed for the detection of refractive index. Examples of those technologies are tapered fibers, FBGs, LPGs, tilted fiber Bragg gratings (TFBGs), Mach-Zehnder interferometers (MZI), single-mode-multimode-single-mode (SMS) structures and also the use of surface plasmon resonance (SPR) excitation [123,124]. POFs present similar properties to those of silica fibers also offering other characteristics, such as flexibility, non-brittle nature, biocompatibility, etc. Furthermore, the possibility to use Bragg gratings in these fibers offers lower complexity on the signal detection, as well as enabling multiplexing, making them attractive to be explored in refractive index sensing.

4.5.1 REFRACTIVE INDEX SENSING WITH FBGs
WRITTEN IN HYGROSCOPIC POFs

As shown before, fibers composed of PMMA have the capability to absorb the water present in the environment. FBGs inscribed in this type of fibers can thus be used to monitor the moisture in the environment by tracking the peak wavelength shift. The phenomenon behind the POFBG wavelength shift is related to the swelling effects of the fiber material as well as the change in its refractive index. If the fiber containing the POFBG is fully immersed in a water container, it will swell, leading to a Bragg wavelength change. The use of this property has been explored on the detection of water dissolved in jet fuel [125,126], which is known to be a thread when the water content nears the fuel's saturation point, leading to hazards associated with: the fuel not burning in an engine; water freeze at high altitudes, plugging fuel filters; microbial growth; corrosion, etc.

The water absorption capabilities presented by PMMA have also been used for concentration sensors. In this case, when a solute is added to the water (where the fiber is immersed) the content of water retained in the polymer matrix will change, leading to a Bragg wavelength shift that depends on the amount of solute present in the solution as a result of the osmosis process [127]. The fiber acts as a semipermeable membrane which blocks the solute but it is permeable to the solvent (water in this example). In such a case, the osmotic pressure created by the solution can be given by Equation 4.24 [127].

$$P_{osm} = CRT \qquad (4.24)$$

where C is the concentration of the solution, R is the gas constant (8.314 J/°C mol), and T is the temperature.

Refractive index characterizations using POFBGs have been performed in salt [47,127,128] and glucose solutions [129]. For that, the POFBGs is submerged in the solution (see the example shown in Figure 4.15a) and enough time is taken to reach the equilibrium and consequently stabilization of the Bragg peak wavelength. The typical Bragg wavelength shift acquired for a POFBG written in a PMMA based SM mPOF for different concentrations of glucose/water concentrations (concentration ranging from 0 to 2.5 mol/L and refractive index ranging from 1.3333 to 1.3955), can be seen in Figure 4.15b [129].

From Figure 4.15b, it can be seen that the resonance Bragg wavelength peak follows a blue-shift with increasing refractive index (solute concentration). The decrease of the Bragg wavelength is related to the amount of solute present in the solution. Therefore, when a solution of a higher concentration covers the POFBG, the osmotic pressure increases, leading to drive the water from the fiber (hypotonic ambient) to the surrounding environment (hypertonic ambient). This will reduce the fiber volume until the PMMA matrix and surrounding solution becomes isotonic. Conversely, this reduction leads to a compression of the Bragg grating period and to a decrease of the effective refractive index, promoting a blue-shift of the Bragg wavelength. Similarly, when the surrounding environment is replaced by a solution of lower concentration the fiber will tend to swell. This will promote the separation between the grating planes and an increase of the effective refractive index, leading to a red-shift of the resonance Bragg wavelength. This type of behavior was

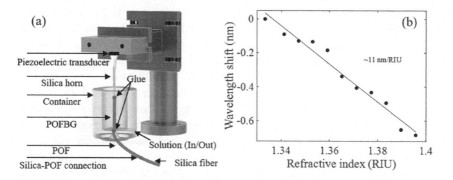

FIGURE 4.15 (a) Schematic of the measurement setup used to characterize the POFBG to different refractive index solutions. (b) Wavelength shift of the peak power Bragg wavelength of a SM-mPOF, for different refractive index of glucose in water solutions.

well described by Zhang et al. [127] where the sensor response for increasing and decreasing concentration of the solution has shown good agreement. Regarding the sensitivity of these fiber optic sensors, values of ~0.37 nm/M [128] and 11 nm/RIU ~ 0.26 nm/M [129] were reported for water/salt solutions and for water/glucose solutions, respectively. Hence, these values compare very favorably with silica FBGs covered with polymer sensitive layers [130–132].

4.5.2   REFRACTIVE INDEX SENSING WITH POF BASED – LPGS, TFBGS, AND FP CAVITIES

The demonstration of gratings in POFs for the detection of the external refractive index does not only rely on the water absorption capabilities of the polymer material. In 2009, David Saéz-Rodrigues et al. demonstrated the capability to measure refractive index by exposing the cladding modes of an LPG written in a PMMA SM-mPOF to different Cargille oils, with refractive index ranging from 1.43 to 1.46 at 532 nm [133]. In their work, it was demonstrated that positive wavelength shifts are observed up to the refractive index of the fiber material ($n \sim 1.49$), while negative wavelength shifts are observed for refractive index higher than the host material. The maximum sensitivity was attained for refractive index values close to the fiber material, which could be interesting when compared with Bragg gratings in silica fibers ($n_{POF}(1.49 @ 532 \text{ nm}) > n_{Silica}(1.46 @ 532 \text{ nm})$). Similar results conducted by another group using sugar/water solutions and an LPG with dip located at the 1,550 nm, written through a 355 nm UV pulsed laser on the commercial available perfluorinated Giga-POF, revealed a blue wavelength shift when the fiber was immersed in water and a red-wavelength shift when the concentration was increased [134].

The use of TFBGs written in POFs has also been reported in literature for the detection of refractive index [135, 136]. The first characterization was reported in a trans-4-stilbenemethanol (TSB) doped SI-PMMA fiber with core and cladding refractive indices at 589 nm of 1.5086 and 1.4904, respectively [135]. A TFBG produced in this fiber with a tilt angle of 3° was immersed in different Cargille refractive index oils, ranging from 1.42 to 1.49, and the transmitted spectral response was acquired right after the immersion (probably to prevent swelling effects related to the water affinity of PMMA). The analysis was made using two techniques, either by measuring the global evolution of the area delimited by the cladding modes or by monitoring the shift of the cladding modes close to the cut-off (point at which the mode is no longer guided and becomes radiated). The first technique revealed a monotonic decreasing trend of the area of the cladding modes with increasing refractive index, achieving maximum sensitivity between 1.45 and 1.48. Regarding the second approach, the selected cladding resonances showed a positive wavelength shift with increasing refractive index, reaching maximum sensitivity in the upper limit of the sensitive region, when the mode reached the cut-off wavelength. The attained maximum sensitivity was about 13 nm/RIU which is on the same order of the sensitivity reported for silica FBGs [137]. One year later, the same group was also able to report the first excitation of surface plasmon waves at 1,550 nm region using a TFBG written in the same POF with 6° tilt angle and coated with a 50 nm gold layer [136]. Surface plasmon resonance was excited with radially polarized modes

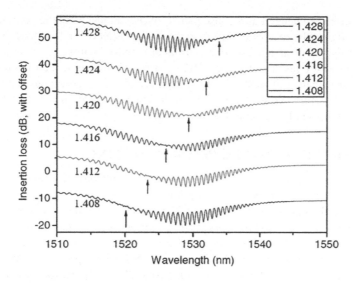

FIGURE 4.16 Evolution of the transmitted spectrum of a 50 nm gold-coated 6° TFBG for different Cargille refractive index oils. (Reprinted from X. Hu, et al., "Surface plasmon excitation at near-infrared wavelengths in polymer optical fibers," *Opt. Lett.*, vol. 40, no. 17, pp. 3998–4001, 2015, with permission from the Optical Society of America.)

(P-polarization, corresponding to TM and EH modes) allowing to tunnel inside the gold coating (electric field is radial at the cladding-gold-coating interface). The SPR signature was obtained at the wavelength of ~1,520 nm for which the phase matching condition is verified. Characterizations of the 50 nm gold-coated POF-TFBG in different refractive index Cargille oils, ranging from 1.428 to 1.408, revealed a redshift of the SPR wavelength (see Figure 4.16) with a linear tendency and sensitivity of ~550 nm/RIU [136].

It is important to mention that when using TFBGs, the range of refractive index values can be tuned by either the tilt angle or the fiber composition. Regarding the lower limit of detection, it is possible to probe smaller refractive indices by increasing the tilt angle of the TFBG because the light is coupled with higher order cladding modes having smaller effective refractive-index values. On the other hand, the increase of the upper limit of detection is restricted by the refractive index of the cladding material. This brings an opportunity for polymer fibers (e.g., PMMA, PC) compared with silica ones. This occurs because the cladding modes will radiate when the refractive index of the external medium approaches the ones of the cladding modes, turning the TFBG insensitive. Thus, considering TFBGs in POFs made of PMMA, the upper limit of detection will be higher than the one achieved for silica fibers (i.e., n_{cl}(POF) ≈ 1.48 > n_{cl}(silica) ≈ 1.44, @ 1,550 nm).

Other fiber optic technologies have also been reported for the refractive index measurement. One of those works has been reported recently for an FP interferometer, in which the cavity was constructed by the use of a short POFBG in a PMMA SM-SI fiber and the Fresnel reflection of the cleaved fiber tip [138]. The measurement principle was based on the change of the fringe visibility when the refractive index of the solutions is changed as a consequence of the change of the reflectivity of the cleaved POF tip. The spectrum of each solution was collected in less than 15 seconds and was analyzed through a fast Fourier transform analysis of the interference signal, obtaining a sensor resolution of 10^{-3} RIU and with little influence on temperature changes. Nevertheless, it should be noted that PMMA is hygroscopic and the long-term water absorption could cause a loss in different wavelength regions [102]; thus, humidity insensitive POFs should be used instead.

4.6 POFBG HYDROSTATIC PRESSURE SENSING

Pressure sensors are important devices in areas ranging from medicine to the oil and gas industry. Monitoring the local variation of pressure allows to prevent leaks in pipelines, which are considered as one of the main threats in the oil and gas industry. Several types of optical fiber sensors have been proposed to measure this parameter using the conventional silica fiber Bragg grating. However, this is a challenging area because of the low compressibility of the silica fiber, which provides inherently low hydrostatic pressure sensitivities (~3–4 pm/MPa) [139,140]. To suppress this problem, the conversion of pressure into strain through special mechanical configurations has been proposed for improving the sensitivity of these fiber optic sensors [141,142]. Yet, the fabrication process could be too elaborate to accomplish.

Polymer optical fibers have received great interest in this subject due to their low Young's modulus and flexibility. In this way, the application of POFBGs for the measurement of hydrostatic pressure has already been explored in a variety of works [14,52,69,70,140,143–146].

4.6.1 THEORY OF FIBER OPTIC PRESSURE SENSORS

When an FBG is under an isotropic stress due to a pressure change (ΔP), the corresponding wavelength shift is given according to Equation 4.2, by:

$$\Delta \lambda_P = \lambda_{\text{Bragg}} \left(\frac{1}{\Lambda} \frac{d\Lambda}{dP} + \frac{1}{n_{\text{eff}}} \frac{dn_{\text{eff}}}{dP} \right) \Delta P \qquad (4.25)$$

This expression has the contribution of two different effects, which are the physical length change $\Delta L/L$, and the refractive index change, that may be expressed as [147]:

$$\frac{\Delta L}{L} = -\frac{(1-2v)P}{E} \qquad (4.26)$$

$$\frac{\Delta n}{n} = \frac{n^2 P}{2E}(1-2v)(2p_{12} + p_{11}) \qquad (4.27)$$

The resultant $\Delta L/L$ is transduced in $\Delta \Lambda/\Lambda$. The normalized pitch pressure and the index pressure coefficients are thus, given by [147,148]:

$$\frac{1}{\Lambda} \frac{d\Lambda}{dP} = -\frac{(1-2v)}{E} \qquad (4.28)$$

$$\frac{1}{n_{\text{eff}}}\frac{dn_{\text{eff}}}{dP} = \frac{n^2}{2E}(1-2v)(2p_{12}+p_{11}) \tag{4.29}$$

The wavelength-pressure sensitivity can be determined by replacing Equations 4.28 and 4.29 into Equation 4.25, producing:

$$S_P = \frac{\lambda_B(1-2v)}{E}\left(\frac{n_{\text{eff}}^2}{2}(2p_{12}+p_{11})-1\right) \tag{4.30}$$

Using the values of silica glass as 70 GPa for the Young's modulus, 0.17 for the Poisson's ration, 0.121 and 0.270 for p_{11} and p_{12}, respectively, an operation wavelength of 1,550 nm and a refractive index of a germanium-doped core of 1.465, gives a theoretical value of −3.3 pm/MPa [139].

From Equation 4.30, it is easy to understand that decreasing Young's modulus the wavelength-pressure sensitivity will increase. This is an advantage for POFBGs since the Young's modulus in polymers is much lower than the one found for silica fibers.

Equation 4.30 is however restricted to isotropic materials like silica. Contrary to silica optical fibers, during the manufacturing process of POFs, the molecular chains tend to align along the length of the fiber [149]. Even using an annealing process, POFs will still show some residual stresses. Therefore, the material cannot be considered isotropic and a more detailed expression needs to be described, which according to Ref. [140] can be written as:

$$S_P = \lambda_B\left[\frac{n_{\text{eff}}^2}{2}\left(\left(\frac{1}{E_P}-\frac{v_P}{E_P}-\frac{v_T}{E_T}\right)(p_{11}+p_{12})+\frac{p_{13}}{E_T}(1-2v_T)\right)-\frac{1}{E_T}(1-2v_T)\right] \tag{4.31}$$

where E_P and E_T are the elastic moduli parallel and perpendicular to the fiber axis, v_P and v_T are the relevant Poisson's ratios, and p_{11}, p_{12}, and p_{13} are the components of the strain-optic tensor [140].

4.6.2 Characterization of POFBG Pressure Sensors

Experimentally, hydrostatic pressure characterizations have been performed using pressurized chambers filled either with air [14,70,140,143] or liquids [52,69,144,145]. However, systems using air are affected by temperature leading to cross sensitive issues. For that reason, the temperature profile of the pressure chamber needs to be measured in order to compensate for the temperature-induced effect on the POFBG [143]. Nevertheless, POFBGs made of materials capable to absorb the water from the environment, such as PMMA [105] and PC [72], will also induce a Bragg wavelength shift when this parameter is changed [62]. Therefore, this parameter needs also to be taken into account in long-term calibration tests. On the other hand, pressure systems using liquids could suppress these effects from the changes in temperature and the amount of water on the environment. Regarding temperature, the incompressible nature of liquids removes this effect due to the rapid pressure changes [145].

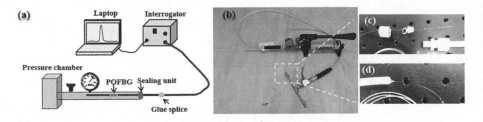

FIGURE 4.17 (a) Schematic example of the experimental hydrostatic pressure character-ization system, composed by the interrogation system and pressure chamber. (b) Example of an oil pressure chamber used for the pressure calibration tests. (c), (d) are the insets of the sealing units used to fix the fiber inside the pressurized pipes through either rubber seal (c) or epoxy resin (d).

Nevertheless, since the fiber is immersed in the liquid, the water surrounding the fiber can be considered constant.

A schematic example of a pressure chamber as well as the POFBG interrogation system may be seen in Figure 4.17a, while an example of an oil pressure chamber may be seen in Figure 4.17b–d.

When inserting the fiber into the pressurized pipe, it is necessary to seal off the zone where the fiber entered, allowing to prevent the gas/liquid from leaking out. For that reason, different authors have used either glue or rubber seal [14,70]. Examples of these type of sealing units are shown in the insets of Figure 4.17c and d. Due to the easiness on single access of the fiber to the pressure chamber, the acquisition of the optical signal is normally made in reflection. The pressure is applied in steps and it is waited enough time for the stabilization of the resonance Bragg peak [69,70,143].

As the pressure increases, the resonance Bragg wavelength is red-shifted, as can be seen in the example shown in Figure 4.18a and b, for the Bragg reflection spec-tra and Bragg wavelength shift, respectively, obtained for a PMMAbased POFBG written at the C-band in a SM mPOF [70].

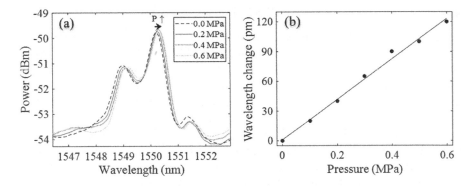

FIGURE 4.18 Spectra (a) and Bragg peak wavelength shift (b), acquired for different pressure conditions for a POFBG recorded in a PMMA mPOF.

When the pressure is increased, the fiber is radially and axially compressed, inducing positive and negative axial strain on the fiber grating planes, respectively. Assuming the same scenario as it occurs with silica fibers, where the axial pressure is the prominent effect [150], it is expected that the net effect is seen as a compression of the grating planes, leading to a negative Bragg wavelength shift. However, the refractive index increases as the fiber is compressed resulting in a positive Bragg wavelength shift [140]. Thus, from the red-shift presented by POFBGs upon an increase of pressure (as shown in Figure 4.18a and b), it is possible to assume that the dominant effect is due to the strain optic effect. This behavior is contrary to the one observed in silica FBGs, in which the physical length change of the period of the grating planes is the most predominant effect, leading to have negative Bragg wavelength shifts with increasing pressure [150].

The first characterization of a POFBG to hydrostatic pressure was made in 2012 by Johnson et al. [140], using a C-band POFBG in a PMMA MM mPOF. Results revealed a hydrostatic pressure sensitivity of ~130 pm/MPa, corresponding to a fractional sensitivity of 83×10^{-6}/MPa, twenty five times higher than the value reported for a silica fiber Bragg grating (i.e., ~ -4 pm/MPa (-2.6×10^{-6}/MPa) [140]). After this preliminary work, several other works followed using different fiber structures, materials, and wavelength regions [14,52,69,70,140,143–146]. The results may be seen in Table 4.5.

As may be seen from Table 4.5, there are significant differences between the different hydrostatic pressure sensitivities reported, even among the same material and wavelength region. The discrepancy between the results may be attributed to the drawing temperature and strain used during the fabrication process [27,151], dopant concentration [27], fiber structure [140,152] and also due to the thermal history after the fiber fabrication [29,153,154]. Such properties can have a strong impact on the mechanical properties of the fibers and thus, on the hydrostatic pressure sensitivity. Nevertheless, it is worth to mention that negative hydrostatic pressure sensitivity results have also been reported for POFBGs [69,146]. Regarding the work presented in Ref. [146], which reports the use of point-by-point (PbP) Bragg grating written by the femtosecond laser on the commercial GigaPOF-50SR, from

TABLE 4.5
Summary of POFBG Hydrostatic Pressure Sensitivity

Fiber	Material	Wavelength (nm)	Sensitivity (pm/MPa)
SM-SI	PMMA	~1,550	200 [143]
SM-mPOF	PMMA	~1,550	204 [70]
SM-mPOF	PMMA	~850	28 [144]
SM-mPOF	BDK-doped PMMA	~850	460 [14]
MM-mPOF	PMMA	~1,550	130 [140]
SM-mPOF	ZEONEX® 480R	~850	30 [145]
MM-GI	CYTOP® + PC	~1,550	−130 & +1,300 [146]
MM-SI	bisphenol-A acrylate	~1,550	−2,800 [69]

Chromis Fiberoptics, it was observed that for short-term measurements (<1 minutes), a blue wavelength shift is observed. According to the authors, this result was due to the greater elongated grating pitch related to the reduced refractive index change. However, for long-term measurements (>2 hours) the Bragg wavelength response was positive, achieving a higher pressure sensitivity of 1,300 pm/MPa. The long-term response was described to be due to the difference in mechanical properties of the materials presented in the polymer core and overcladding.

Concerning high-sensitivity pressure results, it was shown in Ref. [89] that values as high as −2,800 pm/MPa at the 1,550 nm region could be achieved, which is ~700 times higher than the one reported for silica FBGs and the highest among POFBG-based sensors. The fiber was manufactured using a novel light polymerization spinning (LPS) process and using the monomer bisphenol-A acrylate. The grating was written with the PbP method (with 30 μm width, for the selection of the lower order modes [155]), by an fs laser in a large diameter core POF (580 μm). The authors claim that the achievement was due to the unique mechanical properties of the newly presented POF, which exhibit a Young Modulus of 15 MPa which is several orders of magnitude lower than that of POFs made using conventional materials (PMMA 3.3 GPa [97], PC 2.5 GPa [98], TOPAS® 5013 3.2 GPa [100], CYTOP® 1.5 GPa [76], and ZEONEX® 480R 2.2 GPa [99]).

Nevertheless, it is worth to mention that early attempts to achieve higher POFBG hydrostatic pressure sensitivities have been proposed by other authors [143,144]. One of those works reports the use of pre-strained POFBGs to suppress the negative strain induced effect (induced mechanical contraction of the fiber), which is known for its opposite and lower contribution when compared with the stress optic-effect [144]. To do that, the authors used an annealed POFBGs with 850 nm peak wavelength. To pre-strain the POFBG, the fiber was longitudinally strained and fixed with UV glue to an aluminum plate with an amount of strain of 1,100 με. The characterization results revealed that by suppressing the strain-induced effect, it was possible to improve the hydrostatic pressure sensitivity from 28 to 73 pm/MPa [144].

Another way to improve more efficiently, the hydrostatic pressure sensitivity has been accomplished by the reduction of the fiber diameter as described by Bhowmik et al. [30,143,156]. The authors have experimentally demonstrated that the mechanical properties of a POF can change after etching and thus, the pressure sensitivity of POFBGs can be improved. Results have been explored in SM SI-POFs with FBGs at the 1,550 nm region. The fibers were etched during 8 minutes with a mixture of acetone/methanol in the same proportion, obtaining a fiber diameter of 55 μm (initial diameter ~180 μm). Such a diameter reduction allowed the authors to improve the hydrostatic pressure sensitivity from its standard value of 200 pm/MPa, up to 750 pm/MPa [143]. The explanation for their observations was essentially based on the lower Young's modulus of the tapered fiber. Nonetheless, the authors also referred that the tapering process could also affect the Poisson's ratio. However, experimental confirmation on this was not reported.

Beyond POFBG pressure sensors, POF based LPG pressure sensors have also been subject of interest by the research community. One of those works has been performed by Statkiewicz-Barabach et al. [52]. For that, the authors imprinted a 25 mm LPG using transverse periodic loading method together with heat.

The resonance wavelength found at ~700 nm was used to measure the hydrostatic pressure sensitivity, showing a sensitivity as high as 2,300 pm/MPa. This value is much higher than the values commonly found for conventional POFBGs (see Table 4.5) but is still lower than the one recently achieved for POFBG in bisphenol-A acrylate fiber [69].

REFERENCES

1. R. Kashyap, *Fiber Bragg Gratings*. San Diego, CA: Elsevier, 2010.
2. R. M. Waxler, D. Horowitz, and A. Feldman, "Optical and physical parameters of plexiglas 55 and Lexan," *Appl. Opt.*, vol. 18, no. 1, pp. 101–104, 1979.
3. Goodfellow Corporation, "Standard Price List for all Polymers." 2016.
4. T. Ishigure, E. Nihei, and Y. Koike, "Optimum refractive-index profile of the graded-index polymer optical fiber, toward gigabit data links," *Appl. Opt.*, vol. 35, no. 12, pp. 2048–2053, 1996.
5. A. Othonos, "Fiber Bragg gratings," *Rev. Sci. Instrum.*, vol. 68, no. 12, pp. 4309–4341, 1997.
6. S. Kiesel, K. Peters, T. Hassan, and M. Kowalsky, "Behaviour of intrinsic polymer optical fibre sensor for large-strain applications," *Meas. Sci. Technol.*, vol. 18, no. 10, pp. 3144–3154, 2007.
7. P. Antunes, H. Lima, J. Monteiro, and P. S. Andre, "Elastic constant measurement for standard and photosensitive single mode optical fibres," *Microw. Opt. Technol. Lett.*, vol. 50, no. 9, pp. 2467–2469, 2008.
8. Z. Xiong, G. D. Peng, B. Wu, and P. L. Chu, "Highly tunable Bragg gratings in single-mode polymer optical fibers," *IEEE Photonics Technol. Lett.*, vol. 11, no. 3, pp. 352–354, 1999.
9. I.-L. Bundalo, K. Nielsen, G. Woyessa, and O. Bang, "Long-term strain response of polymer optical fiber FBG sensors," *Opt. Mater. Express*, vol. 7, no. 3, pp. 967–976, 2017.
10. A. Stefani, C. Markos, and O. Bang, "Narrow bandwidth 850-nm fiber Bragg gratings in few-mode polymer optical fibers," *IEEE Photonics Technol. Lett.*, vol. 23, no. 10, pp. 660–662, 2011.
11. Z. F. Zhang, C. Zhang, X. M. Tao, G. F. Wang, and G. D. Peng, "Inscription of polymer optical fiber Bragg grating at 962 nm and its potential in strain sensing," *IEEE Photonics Technol. Lett.*, vol. 22, no. 21, pp. 1562–1564, 2010.
12. H. B. Liu, H. Y. Liu, G. D. Peng, and P. L. Chu, "Strain and temperature sensor using a combination of polymer and silica fibre Bragg gratings," *Opt. Commun.*, vol. 219, no. 1–6, pp. 139–142, 2003.
13. W. Yuan et al., "Improved thermal and strain performance of annealed polymer optical fiber Bragg gratings," *Opt. Commun.*, vol. 284, no. 1, pp. 176–182, 2011.
14. L. Pereira et al., "Phase-shifted Bragg grating inscription in PMMA microstructured POF using 248 nm UV radiation," *J. Light. Technol.*, vol. 35, no. 23, pp. 5176–5184, 2017.
15. A. Fasano et al., "Fabrication and characterization of polycarbonate microstructured polymer optical fibers for high-temperature-resistant fiber Bragg grating strain sensors," *Opt. Mater. Express*, vol. 6, no. 2, pp. 649–659, 2016.
16. W. Yuan et al., "Humidity insensitive TOPAS polymer fiber Bragg grating sensor," *Opt. Express*, vol. 19, no. 20, pp. 19731–19739, 2011.
17. C. Markos, A. Stefani, K. Nielsen, H. K. Rasmussen, W. Yuan, and O. Bang, "High-Tg TOPAS microstructured polymer optical fiber for fiber Bragg grating strain sensing at 110 degrees," *Opt. Express*, vol. 21, no. 4, pp. 4758–4765, 2013.

18. R. Oliveira, T. H. R. Marques, L. Bilro, R. Nogueira, and C. M. B. Cordeiro, "Multiparameter POF sensing based on multimode interference and fiber Bragg grating," *J. Light. Technol.*, vol. 35, no. 1, pp. 3–9, 2017.

19. G. Woyessa, A. Fasano, C. Markos, A. Stefani, H. K. Rasmussen, and O. Bang, "Zeonex microstructured polymer optical fiber: Fabrication friendly fibers for high temperature and humidity insensitive Bragg grating sensing," *Opt. Mater. Express*, vol. 7, no. 1, pp. 286–295, 2017.

20. R. Min, B. Ortega, A. Leal-Junior, and C. Marques, "Fabrication and characterization of Bragg grating in CYTOP POF at 600-nm wavelength," *IEEE Sensors Lett.*, vol. 2, no. 3, pp. 1–4, 2018.

21. Y. Zheng, K. Bremer, and B. Roth, "Investigating the strain, temperature and humidity sensitivity of a multimode graded-index perfluorinated polymer optical fiber with Bragg grating," *Sensors*, vol. 18, no. 5, p. 1436, 2018.

22. R. Ishikawa et al., "Strain dependence of perfluorinated polymer optical fiber Bragg grating measured at different wavelengths," *Jpn. J. Appl. Phys.*, vol. 57, no. 3, p. 038002, 2018.

23. H. Y. Liu, H. B. Liu, and G. D. Peng, "Tensile strain characterization of polymer optical fibre Bragg gratings," *Opt. Commun.*, vol. 251, no. 1–3, pp. 37–43, 2005.

24. G. Statkiewicz-Barabach, P. Mergo, and W. Urbanczyk, "Bragg grating-based Fabry–Perot interferometer fabricated in a polymer fiber for sensing with improved resolution," *J. Opt.*, vol. 19, no. 1, p. 015609, 2017.

25. J. A. Tucknott, L. Reekie, L. Dong, M. G. Xu, and J. L. Cruz, "Temperature-independent strain sensor using a chirped Bragg grating in a tapered optical fibre," *Electron. Lett.*, vol. 31, no. 10, pp. 823–825, 1995.

26. R. Min et al., "Microstructured PMMA POF chirped Bragg gratings for strain sensing," *Opt. Fiber Technol.*, vol. 45, pp. 330–335, 2018.

27. C. Jiang, M. G. Kuzyk, J.-L. Ding, W. E. Johns, and D. J. Welker, "Fabrication and mechanical behavior of dye-doped polymer optical fiber," *J. Appl. Phys.*, vol. 92, no. 1, pp. 4–12, 2002.

28. A. Pospori et al., "Annealing effects on strain and stress sensitivity of polymer optical fibre based sensors," in *Micro-Structured and Specialty Optical Fibres IV*, K. Kalli and A. Mendez, Eds. Bellingham: SPIE, 2016, vol. 9886, p. 98860V.

29. A. Pospori, C. A. F. Marques, D. Sáez-Rodríguez, K. Nielsen, O. Bang, and D. J. Webb, "Thermal and chemical treatment of polymer optical fiber Bragg grating sensors for enhanced mechanical sensitivity," *Opt. Fiber Technol.*, vol. 36, pp. 68–74, 2017.

30. K. Bhowmik et al., "Intrinsic high-sensitivity sensors based on etched single-mode polymer optical fibers," *IEEE Photonics Technol. Lett.*, vol. 27, no. 6, pp. 604–607, 2015.

31. A. Abang and D. J. Webb, "Effects of annealing, pre-tension and mounting on the hysteresis of polymer strain sensors," *Meas. Sci. Technol.*, vol. 25, no. 1, p. 015102, 2014.

32. A. Abang and D. J. Webb, "Influence of mounting on the hysteresis of polymer fiber Bragg grating strain sensors," *Opt. Lett.*, vol. 38, no. 9, pp. 1376–1378, 2013.

33. M. C. J. Large, J. Moran, and L. Ye, "The role of viscoelastic properties in strain testing using microstructured polymer optical fibres (mPOF)," *Meas. Sci. Technol.*, vol. 20, no. 3, p. 034014, 2009.

34. A. Stefani, S. Andresen, W. Yuan, and O. Bang, "Dynamic characterization of polymer optical fibers," *IEEE Sens. J.*, vol. 12, no. 10, pp. 3047–3053, 2012.

35. W. Yuan, A. Stefani, and O. Bang, "Tunable polymer fiber Bragg grating (FBG) inscription: Fabrication of dual-FBG temperature compensated polymer optical fiber strain sensors," *Photonics Technol. Lett.*, vol. 24, no. 5, pp. 401–403, 2012.

36. A. Leal-Junior et al., "Dynamic mechanical characterization with respect to temperature, humidity, frequency and strain in mPOFs made of different materials," *Opt. Mater. Express*, vol. 8, no. 4, pp. 804–815, 2018.

37. D. Sáez-Rodríguez, K. Nielsen, O. Bang, and D. J. Webb, "Time-dependent variation of fibre Bragg grating reflectivity in PMMA based polymer optical fibres," *Opt. Lett.*, vol. 40, no. 7, pp. 1476–1479, 2015.

38. J. Rösler, H. Harders, and M. Bäker, *Mechanical Behaviour of Engineering Materials.* New York: Springer-Verlag, 2007.

39. A. Stefani, S. Andresen, W. Yuan, N. Herholdt-Rasmussen, and O. Bang, "High sensitivity polymer optical fiber-Bragg-grating-based accelerometer," *IEEE Photonics Technol. Lett.*, vol. 24, no. 9, pp. 763–765, 2012.

40. A. Stefani et al., "Temperature compensated, humidity insensitive, high-Tg TOPAS FBGs for accelerometers and microphones," in *22nd International Conference on Optical Fiber Sensors*, Y. Liao et al., Eds. Bellingham: SPIE, 2012, vol. 8421, p. 84210Y.

41. R. Oliveira, T. H. R. Marques, and C. M. B. Cordeiro, "Strain sensitivity enhancement of a sensing head based on ZEONEX polymer FBG in series with silica fiber," *J. Light. Technol.*, vol. 36, no. 22, pp. 5106–5112, 2018.

42. R. Oliveira, L. Bilro, and R. Nogueira, "Strain sensitivity control of an in-series silica and polymer FBG," *Sensors*, vol. 18, no. 6, p. 1884, 2018.

43. T. X. Wang, Y. H. Luo, G. D. Peng, and Q. J. Zhang, "High-sensitivity stress sensor based on Bragg grating in BDK-doped photosensitive polymer optical fiber," in *Third Asia Pacific Optical Sensors Conference*, J. Canning and G.-D. Peng, Eds. Bellingham: SPIE, 2012, vol. 8351, p. 83510M.

44. Y. Luo et al., "Analysis of multimode POF gratings in stress and strain sensing applications," *Opt. Fiber Technol.*, vol. 17, no. 3, pp. 201–209, 2011.

45. E. R. Lyons and H. P. Lee, "Demonstration of an etched cladding fiber Bragg grating filter with reduced tuning force requirement," *IEEE Photonics Technol. Lett.*, vol. 11, no. 12, pp. 1626–1628, 1999.

46. G. Rajan, B. Liu, Y. Luo, E. Ambikairajah, and G.-D. Peng, "High sensitivity force and pressure measurements using etched singlemode polymer fiber Bragg gratings," *IEEE Sens. J.*, vol. 13, no. 5, pp. 1794–1800, 2013.

47. X. Hu, C.-F. J. Pun, H.-Y. Tam, P. Mégret, and C. Caucheteur, "Highly reflective Bragg gratings in slightly etched step-index polymer optical fiber," *Opt. Express*, vol. 22, no. 15, pp. 18807–18817, 2014.

48. K. Bhowmik et al., "Etching process related changes and effects on solid-core singlemode polymer optical fiber grating," *IEEE Photonics J.*, vol. 8, no. 1, p. 2500109, 2016.

49. O. Frazão, S. F. O. Silva, A. Guerreiro, J. L. Santos, L. A. Ferreira, and F. M. Araújo, "Strain sensitivity control of fiber Bragg grating structures with fused tapers," *Appl. Opt.*, vol. 46, no. 36, pp. 8578–8582, 2007.

50. G. Durana, J. Gomez, G. Aldabaldetreku, J. Zubia, A. Montero, and I. S. de Ocariz, "Assessment of an LPG mPOF for strain sensing," *IEEE Sens. J.*, vol. 12, no. 8, pp. 2668–2673, 2012.

51. K. Krebber, P. Lenke, S. Liehr, J. Witt, and M. Schukar, "Smart technical textiles with integrated POF sensors," in *Smart Sensor Phenomena, Technology, Networks, and Systems*, W. Ecke, K. J. Peters, N. G. Meyendorf, Eds. Bellingham: SPIE, 2008, vol. 6933, p. 69330V.

52. G. Statkiewicz-Barabach, D. Kowal, M. K. Szczurowski, P. Mergo, and W. Urbanczyk, "Hydrostatic pressure and strain sensitivity of long period grating fabricated in polymer microstructured fiber," *IEEE Photonics Technol. Lett.*, vol. 25, no. 5, pp. 496–499, 2013.

53. M. Steffen, M. Schukar, J. Witt, K. Krebber, M. Large, and A. Argyros, "Investigation of mPOF-LPGs for sensing applications," in *18th International Conference on Plastic Optical Fibers*, 2009, pp. 25–26.

54. J. Witt, M. Steffen, M. Schukar, and K. Krebber, "Investigation of sensing properties of long period gratings based on microstructured polymer optical fibres," in *Fourth European Workshop on Optical Fibre Sensors*, J. L. Santos et al., Eds. Bellingham: SPIE, 2010, vol. 7653, p. 76530I.

55. R. Lwin, A. Argyros, S. G. Leon-Saval, and M. C. J. Large, "Strain sensing using long period gratings in microstructured polymer optical fibres," in *21st International Conference on Optical Fiber Sensors*, W. J. Bock, J. Albert, and X. Bao, Eds. Bellingham: SPIE, 2011, vol. 7753, p. 775396.

56. V. Bhatia et al., "Temperature-insensitive and strain-insensitive long-period grating sensors for smart structures," *Opt. Eng.*, vol. 36, no. 7, pp. 1872–1876, 1997.

57. D. Kowal, G. Statkiewicz-Barabach, P. Mergo, and W. Urbanczyk, "Inscription of long period gratings using an ultraviolet laser beam in the diffusion-doped microstructured polymer optical fiber," *Appl. Opt.*, vol. 54, no. 20, pp. 6327–6333, 2015.

58. I.-L. Bundalo, R. Lwin, S. Leon-Saval, and A. Argyros, "All-plastic fiber-based pressure sensor," *Appl. Opt.*, vol. 55, no. 4, pp. 811–816, 2016.

59. R. Min, C. A. F. Marques, K. Nielsen, O. Bang, and B. Ortega, "Fast inscription of long period gratings in microstructured polymer optical fibers," *IEEE Sens. J.*, vol. 18, no. 5, pp. 1919–1923, 2018.

60. G. D. Peng, Z. Xiong, and P. L. Chu, "Photosensitivity and gratings in dye-doped polymer optical fibers," *Opt. Fiber Technol.*, vol. 5, no. 2, pp. 242–251, 1999.

61. H. Y. Liu, G. D. Peng, and P. L. Chu, "Thermal tuning of polymer optical fiber Bragg gratings," *IEEE Photonics Technol. Lett.*, vol. 13, no. 8, pp. 824–826, 2001.

62. N. Harbach, *Fiber Bragg Gratings in Polymer Optical Fibers*. Lausanne: École Polytechnique Fédérale de Lausanne, 2008.

63. Z. F. Zhang and X. Ming Tao, "Synergetic effects of humidity and temperature on PMMA based fiber Bragg gratings," *J. Light. Technol.*, vol. 30, no. 6, pp. 841–845, 2012.

64. G. N. Harbach, H. G. Limberger, and R. P. Salathé, "Influence of humidity and temperature on polymer optical fiber Bragg gratings," in *Advanced Photonics & Renewable Energy*, 2010, p. BTuB2.

65. K. E. Carroll, C. Zhang, D. J. Webb, K. Kalli, A. Argyros, and M. C. Large, "Thermal response of Bragg gratings in PMMA microstructured optical fibers," *Opt. Express*, vol. 15, no. 14, pp. 8844–8850, 2007.

66. K. Minami, *Handbook of Plastic Optics*, 2nd ed. Weinheim, Germany: Wiley-VCH Verlag GmbH & Co. KGaA, 2010.

67. A. Lacraz, M. Polis, A. Theodosiou, C. Koutsides, and K. Kalli, "Bragg grating inscription in CYTOP polymer optical fibre using a femtosecond laser," in *Micro-Structured and Specialty Optical Fibres IV*, K. Kalli, J. Kanka and A. Mendez, Eds. Bellingham: SPIE, 2015, vol. 9507, p. 95070K.

68. A. Theodosiou et al., "Measurements with an FBG inscribed on a new type of polymer fibre," in *26th International Conference on Plastic Optical Fibres*, 2017, vol. 26, p. 74.

69. A. Leal-Junior et al., "Characterization of a new polymer optical fiber with enhanced sensing capabilities using a Bragg grating," *Opt. Lett.*, vol. 43, no. 19, pp. 4799–4802, 2018.

70. R. Nogueira, R. Oliveira, L. Bilro, and J. Heidarialamdarloo, "New advances in polymer fiber Bragg gratings," *Opt. Laser Technol.*, vol. 78, pp. 104–109, 2016.

71. X. Cheng et al., "High-sensitivity temperature sensor based on Bragg grating in BDK-doped photosensitive polymer optical fiber," *Chinese Opt. Lett.*, vol. 9, no. 2, p. 020602, 2011.

72. G. Woyessa, A. Fasano, C. Markos, H. Rasmussen, and O. Bang, "Low loss polycarbonate polymer optical fiber for high temperature FBG humidity sensing," *IEEE Photonics Technol. Lett.*, vol. 29, no. 7, pp. 575–578, 2017.

73. I. P. Johnson et al., "Optical fibre Bragg grating recorded in TOPAS cyclic olefin copolymer," *Electron. Lett.*, vol. 47, no. 4, pp. 271–272, 2011.

74. G. Woyessa et al., "Zeonex-PMMA microstructured polymer optical FBGs for simultaneous humidity and temperature sensing," *Opt. Lett.*, vol. 42, no. 6, pp. 1161–1164, 2017.

75. A. Theodosiou, X. Hu, C. Caucheteur, and K. Kalli, "Bragg gratings and Fabry-Perot cavities in low-loss multimode CYTOP polymer fiber," *IEEE Photonics Technol. Lett.*, vol. 30, no. 9, pp. 857–860, 2018.

76. Asahi Glass Co. Ltd., "Amorphous Fluoropolymer (CYTOP)." Tokyo, 2009.

77. R. Oliveira, C. A. F. Marques, L. Bilro, and R. N. Nogueira, "Production and characterization of Bragg gratings in polymer optical fibers for sensors and optical communications," in *23rd International Conference on Optical Fiber Sensors*, J. M. López-Higuera et al., Eds. Bellingham: SPIE, 2014, vol. 9157, p. 915794.

78. W. Zhang and D. J. Webb, "Factors influencing the temperature sensitivity of PMMA based optical fiber Bragg gratings," in *Micro-structured and Specialty Optical Fibres III*, K. Kalli and A. Mendez, Eds. Bellingham: SPIE, 2014, vol. 9128, p. 91280M.

79. K. Bhowmik et al., "High intrinsic sensitivity etched polymer fiber Bragg grating pair for simultaneous strain and temperature measurements," *IEEE Sens. J.*, vol. 16, no. 8, pp. 2453–2459, 2016.

80. K. Kalli et al., "Electrically tunable Bragg gratings in single-mode polymer optical fiber," *Opt. Lett.*, vol. 32, no. 3, pp. 214–216, 2007.

81. T. L. Yeo, T. Sun, and K. T. V Grattan, "Fibre-optic sensor technologies for humidity and moisture measurement," *Sensors Actuators A Phys.*, vol. 144, no. 2, pp. 280–295, 2008.

82. T. L. Yeo, K. T. V. Grattan, D. Parry, R. Lade, and B. D. Powell, "Polymer-coated fiber Bragg grating for relative humidity sensing," *IEEE Sens. J.*, vol. 5, no. 5, pp. 1082–1089, 2005.

83. M. Ams et al., "Fibre optic temperature and humidity sensors for harsh wastewater environments," in *Eleventh International Conference on Sensing Technology (ICST)*, 2017, pp. 1–3.

84. J. S. Santos, I. M. Raimundo, C. M. B. Cordeiro, C. R. Biazoli, C. A. J. Gouveia, and P. A. S. Jorge, "Characterisation of a Nafion film by optical fibre Fabry-Perot interferometry for humidity sensing," *Sensors Actuators B Chem.*, vol. 196, pp. 99–105, 2014.

85. L. H. Chen et al., "Chemical Chitosan based fiber-optic Fabry-Perot humidity sensor," *Sensors Actuators B Chem.*, vol. 169, pp. 167–172, 2012.

86. T. Venugopalan, T. L. Yeo, T. Sun, and K. T. V. Grattan, "LPG-based PVA coated sensor for relative humidity measurement," *IEEE Sens. J.*, vol. 8, no. 7, pp. 1093–1098, 2008.

87. H. Sun, X. Zhang, L. Yuan, L. Zhou, X. Qiao, and M. Hu, "An optical fiber Fabry-Perot interferometer sensor for simultaneous measurement of relative humidity and temperature," *IEEE Sens. J.*, vol. 15, no. 5, pp. 2891–2897, 2015.

88. H. Liu, H. Liang, M. Sun, K. Ni, and Y. Jin, "Simultaneous measurement of humidity and temperature based on a long-period fiber grating inscribed in fiber loop mirror," *IEEE Sens. J.*, vol. 14, no. 3, pp. 893–896, 2014.

89. S. Zhang, X. Dong, T. Li, C. C. Chan, and P. P. Shum, "Simultaneous measurement of relative humidity and temperature with PCF-MZI cascaded by fiber Bragg grating," *Opt. Commun.*, vol. 303, pp. 42–45, 2013.

90. J. Mathew, Y. Semenova, G. Rajan, P. Wang, and G. Farrell, "Improving the sensitivity of a humidity sensor based on fiber bend coated with a hygroscopic coating," *Opt. Laser Technol.*, vol. 43, no. 7, pp. 1301–1305, 2011.
91. C. Massaroni, M. Caponero, R. D'Amato, D. Lo Presti, and E. Schena, "Fiber Bragg grating measuring system for simultaneous monitoring of temperature and humidity in mechanical ventilation," *Sensors*, vol. 17, no. 4, p. 749, 2017.
92. C. Massaroni, D. Lo Presti, P. Saccomandi, M. A. Caponero, R. D'Amato, and E. Schena, "Fiber Bragg grating probe for relative humidity and respiratory frequency estimation: Assessment during mechanical ventilation," *IEEE Sens. J.*, vol. 18, no. 5, pp. 2125–2130, 2018.
93. C. Lee, Y. You, J. Dai, J. Hsu, and J. Horng, "Hygroscopic polymer microcavity fiber Fizeau interferometer incorporating a fiber Bragg grating for simultaneously sensing humidity and temperature," *Sensors Actuators B Chem.*, vol. 222, pp. 339–346, 2016.
94. C.-T. Ma, Y.-W. Chang, Y.-J. Yang, and C.-L. Lee, "A dual-polymer fiber Fizeau interferometer for simultaneous measurement of relative humidity and temperature," *Sensors*, vol. 17, no. 11, p. 2659, 2017.
95. D. Viegas et al., "Simultaneous measurement of humidity and temperature based on an SiO2-nanospheres film deposited on a long-period grating in-line with a fiber Bragg grating," *IEEE Sens. J.*, vol. 11, no. 1, pp. 162–166, 2011.
96. Y. Luo, B. Yan, Q. Zhang, G.-D. Peng, J. Wen, and J. Zhang, "Fabrication of polymer optical fibre (POF) gratings," *Sensors*, vol. 17, no. 511, 2017.
97. Mitsubishi Rayon, "General Properties of AcrypetTM." 2015.
98. Teijin Limited, "Panlite® AD-5503- Polycarbonate." 2016.
99. Zeon Chemicals, "ZEONEX® Cyclo Olefin Polymer (COP)." 2016.
100. Topas Advanced Polymers GmbH, "Cycloolefin Copolymer (COC)." 2013.
101. G. Khanarian and H. Celanese, "Optical properties of cyclic olefin copolymers," *Opt. Eng.*, vol. 40, no. 6, pp. 1024–1029, 2001.
102. T. Kaino, "Influence of water absorption on plastic optical fibers," *Appl. Opt.*, vol. 24, no. 23, pp. 4192–4195, 1985.
103. J. Witt, M. Breithaupt, J. Erdmann, and K. Krebber, "Humidity sensing based on microstructured POF long period gratings," in *20th International Conference on Plastic Optical Fibers*, 2011, pp. 409–414.
104. R. Oliveira, T. H. R. Marques, L. Bilro, C. M. B. Cordeiro, and R. N. Nogueira, "Strain, temperature and humidity sensing with multimode interference in POF," in *25th International Conference on Plastic Optical Fibers*, 2016, p. OP35.
105. G. Woyessa, K. Nielsen, A. Stefani, C. Markos, and O. Bang, "Temperature insensitive hysteresis free highly sensitive polymer optical fiber Bragg grating humidity sensor," *Opt. Express*, vol. 24, no. 2, pp. 1206–1213, 2016.
106. C. Zhang, W. Zhang, D. J. Webb, and G.-D. Peng, "Optical fibre temperature and humidity sensor," *Electron. Lett.*, vol. 46, no. 9, pp. 643–644, 2010.
107. W. Zhang, D. J. Webb, and G.-D. Peng, "Investigation into time response of polymer fiber Bragg grating based humidity sensors," *J. Light. Technol.*, vol. 30, no. 8, pp. 1090–1096, 2012.
108. X. Chen, W. Zhang, C. Liu, Y. Hong, and D. J. Webb, "Enhancing the humidity response time of polymer optical fiber Bragg grating by using laser micromachining," *Opt. Express*, vol. 23, no. 20, pp. 25942–25949, 2015.
109. W. Zhang and D. J. Webb, "Humidity responsivity of poly(methyl methacrylate)-based optical fiber Bragg grating sensors," *Opt. Lett.*, vol. 39, no. 10, pp. 3026–3029, 2014.
110. T. Watanabe, N. Ooba, Y. Hida, and M. Hikita, "Influence of humidity on refractive index of polymers for optical waveguide and its temperature dependence," *Appl. Phys. Lett.*, vol. 72, no. 13, pp. 1533–1535, 1998.

111. R. Oliveira, L. Bilro, T. H. R. Marques, C. M. B. Cordeiro, and R. Nogueira, "Simultaneous detection of humidity and temperature through an adhesive based Fabry-Pérot cavity combined with polymer fiber Bragg grating," *Opt. Lasers Eng.*, vol. 114, pp. 37–43, 2019.

112. J. M. Cariou, J. Dugas, L. Martin, and P. Michel, "Refractive-index variations with temperature of PMMA and polycarbonate," *Appl. Opt.*, vol. 25, no. 3, pp. 334–336, 1986.

113. A. D. Kersey et al., "Fiber grating sensors," *J. Light. Technol.*, vol. 15, no. 8, pp. 1442–1463, 1997.

114. K. J. Kim, A. Bar-Cohen, and B. Han, "Thermo-optical modeling of polymer fiber Bragg grating illuminated by light emitting diode," *Int. J. Heat Mass Transf.*, vol. 50, no. 25–26, pp. 5241–5248, 2007.

115. J. Bonefacino et al., "Ultra-fast polymer optical fibre Bragg grating inscription for medical devices," *Light Sci. Appl.*, vol. 7, no. 3, p. 17161, 2018.

116. G. Woyessa, J. K. M. Pedersen, A. Fasano, K. Nielsen, C. Markos, H. K. Rasmussen, and O. Bang, "Simultaneous measurement of temperature and humidity with microstructured polymer optical fiber Bragg gratings," in *25th International Conference on Optical Fiber Sensors*, Y. Chung et al., Eds. Bellingham: SPIE, 2017, vol. 10323, p. 103234.

117. G. Woyessa, A. Fasano, A. Stefani, C. Markos, H. K. Rasmussen, and O. Bang, "Single mode step-index polymer optical fiber for humidity insensitive high temperature fiber Bragg grating sensors," *Opt. Express*, vol. 24, no. 2, pp. 1253–1260, 2016.

118. J. Bonefacino, M.-L. V. Tse, X. Cheng, C.-F. J. Pun, and H.-Y. Tam, "Reliability of PMMA-based FBG for humidity sensing," in *Novel Optical Materials and Applications*, 2016, p. JTu4A.31.

119. G. Rajan, Y. M. Noor, B. Liu, E. Ambikairaja, D. J. Webb, and G.-D. Peng, "A fast response intrinsic humidity sensor based on an etched singlemode polymer fiber Bragg grating," *Sensors Actuators A Phys.*, vol. 203, pp. 107–111, 2013.

120. J. Crank, *The Mathematics of Diffusion*, 2nd ed. Oxford: Clarendon Press, 1975.

121. D. T. Turner, "Polymethyl methacrylate plus water: sorption kinetics and volumetric changes," *Polymer*, vol. 23, no. 2, pp. 197–202, 1982.

122. D. Saez-Rodriguez, J. L. Cruz, I. Johnson, D. J. Webb, M. C. J. Large, and A. Argyros, "Water diffusion into UV inscripted long period grating in microstructured polymer fiber," *IEEE Sens. J.*, vol. 10, no. 7, pp. 1169–1173, 2010.

123. P. A. S. Jorge et al., "Fiber optic-based refractive index sensing at INESC porto," *Sensors*, vol. 12, no. 6, pp. 8371–8389, 2012.

124. S. Silva, P. Roriz, and O. Frazão, "Refractive index measurement of liquids based on microstructured optical fibers," *Photonics*, vol. 1, no. 4, pp. 516–529, 2014.

125. C. Zhang, X. Chen, D. J. Webb, and G.-D. Peng, "Water detection in jet fuel using a polymer optical fibre Bragg grating," in *20th International Conference on Optical Fibre Sensors*, J. Jones et al., Eds. Bellingham: SPIE, 2009, vol. 7503, p. 750380.

126. W. Zhang, D. J. Webb, M. Carpenter, and C. Williams, "Measuring water activity of aviation fuel using a polymer optical fiber Bragg grating," in *23rd International Conference on Optical Fibre Sensors*, J. M. López-Higuera et al., Eds. Bellingham: SPIE, 2014, vol. 9157, p. 91574V.

127. W. Zhang, D. J. Webb, and G.-D. Peng, "Polymer optical fiber Bragg grating acting as an intrinsic biochemical concentration sensor," *Opt. Lett.*, vol. 37, no. 8, pp. 1370–1372, 2012.

128. Z. Zhang, *Bragg Grating Formation in PMMA Fibers Doped with Trans-4-Stilbenemethanol*. Hong Kong: The Hong Kong Polytechnic University, 2013.

129. R. Ferreira, L. Bilro, C. Marques, R. Oliveira, and R. Nogueira, "Refractive index and viscosity: Dual sensing with plastic fibre gratings," in *23rd International Conference on Optical Fibre Sensors*, J. M. López-Higuera et al., Eds. Bellingham: SPIE, 2014, vol. 9157, p. 915793.

130. J. Cong, X. Zhang, K. Chen, and J. Xu, "Fiber optic Bragg grating sensor based on hydrogels for measuring salinity," *Sensors Actuators B Chem.*, vol. 87, no. 3, pp. 487–490, 2002.

131. L. Men, P. Lu, and Q. Chen, "A multiplexed fiber Bragg grating sensor for simultaneous salinity and temperature measurement," *J. Appl. Phys.*, vol. 103, no. 5, p. 053107, 2008.

132. P. Lu, L. Men, and Q. Chen, "Polymer-coated fiber bragg grating sensors for simultaneous monitoring of soluble analytes and temperature," *IEEE Sens. J.*, vol. 9, no. 4, pp. 340–345, 2009.

133. D. Sáez-Rodríguez, J. L. C. Munoz, I. Johnson, D. J. Webb, M. C. J. Large, and A. Argyros, "Long period fibre gratings photoinscribed in a microstructured polymer optical fibre by UV radiation," in *Photonic Crystal Fibers III*, K. Kalli, Ed. Bellingham: SPIE, 2009, vol. 7357, p. 73570L.

134. J. Castrellon-Uribe, M. Lomer, H. Roufael, L. Rodriguez-Cobo, and J. M. Lopez-Higuera, "LPG in perfluorinated GI-POF for concentration measurement in liquids," in *Advanced Photonics 2013*, 2013, p. JT3A.27.

135. X. Hu, C.-F. J. Pun, H.-Y. Tam, P. Mégret, and C. Caucheteur, "Tilted Bragg gratings in step-index polymer optical fiber," *Opt. Lett.*, vol. 39, no. 24, pp. 6835–6838, 2014.

136. X. Hu, P. Mégret, and C. Caucheteur, "Surface plasmon excitation at near-infrared wavelengths in polymer optical fibers," *Opt. Lett.*, vol. 40, no. 17, pp. 3998–4001, 2015.

137. C. Chan, C. Chen, A. Jafari, A. Laronche, D. J. Thomson, and J. Albert, "Optical fiber refractometer using narrowband cladding-mode resonance shifts," *Appl. Opt.*, vol. 46, no. 7, pp. 1142–1149, 2007.

138. M. F. S. Ferreira, G. Statkiewicz-Barabach, D. Kowal, P. Mergo, W. Urbanczyk, and O. Frazão, "Fabry-Perot cavity based on polymer FBG as refractive index sensor," *Opt. Commun.*, vol. 394, pp. 37–40, 2017.

139. M. G. Xu, L. Reekie, Y. T. Chow, and J. P. Dakin, "Optical in-fibre grating high pressure sensor," *Electron. Lett.*, vol. 29, no. 4, pp. 398–399, 1993.

140. I. P. Johnson, D. J. Webb, and K. Kalli, "Hydrostatic pressure sensing using a polymer optical fibre Bragg gratings," in *Third Asia Pacific Optical Sensors Conference*, J. Canning and G.-D. Peng, Eds. Bellingham: SPIE, 2012, vol. 8351, p. 835106.

141. W. T. Zhang, F. Li, Y. L. Liu, and L. H. Liu, "Ultrathin FBG pressure sensor with enhanced responsivity," *IEEE Photonics Technol. Lett.*, vol. 19, no. 19, pp. 1553–1555, 2007.

142. Ying Zhang et al., "High-sensitivity pressure sensor using a shielded polymer-coated fiber Bragg grating," *IEEE Photonics Technol. Lett.*, vol. 13, no. 6, pp. 618–619, 2001.

143. K. Bhowmik et al., "Experimental study and analysis of hydrostatic pressure sensitivity of polymer fibre Bragg gratings," *J. Light. Technol.*, vol. 33, no. 12, pp. 2456–2462, 2015.

144. J. K. M. Pedersen, G. Woyessa, K. Nielsen, and O. Bang, "Effects of pre-strain on the intrinsic pressure sensitivity of polymer optical fiber Bragg gratings," in *25th International Conference on Optical Fiber Sensors*, Y. Chung et al., Eds. Bellingham: SPIE, 2017, vol. 10323, p. 103234U.

145. J. K. M. Pedersen, G. Woyessa, K. Nielsen, and O. Bang, "Intrinsic pressure response of a single-mode cyclo olefin polymer microstructured optical fibre Bragg grating," *25th International Conference on Plastic Optical Fibers*, 2016, p. PP29.

146. R. Ishikawa et al., "Pressure dependence of fiber Bragg grating inscribed in perfluorinated plastic optical fiber using femtosecond laser," *IEEE Photonics Technol. Lett.*, vol. 29, no. 24, pp. 2167–2170, 2017.

147. G. B. Hocker, "Fiber-optic sensing of pressure and temperature," *Appl. Opt.*, vol. 18, no. 9, pp. 1445–1448, 1979.

148. Y.-J. Rao, "In-fibre Bragg grating sensors," *Meas. Sci. Technol.*, vol. 8, no. 4, pp. 355–375, 1997.

149. P. Ji, A. D. Q. Li, and G.-D. Peng, "Transverse birefringence in polymer optical fiber introduced in drawing process," in *Linear and Nonlinear Optics of Organic Materials III*, M. G. Kuzyk, M. Eich and R. A. Norwood, Eds. Bellingham: SPIE, 2003, vol. 5212, pp. 108–117.

150. C. Wu, B. Guan, Z. Wang, and X. Feng, "Characterization of pressure response of Bragg gratings in grapefruit microstructured fibers," *J. Light. Technol.*, vol. 28, no. 9, pp. 1392–1397, 2010.

151. M. K. Szczurowski et al., "Measurements of stress-optic coefficient and Young's modulus in PMMA fibers drawn under different conditions," in *Photonic Crystal Fibres IV*, K. Kalli and W. Urbanczyk, Eds. Bellingham: SPIE, 2010, vol. 7714, p. 77140G.

152. A. G. Leal-Junior et al., "Influence of the cladding structure in PMMA mPOFs mechanical properties for strain sensors applications," *IEEE Sens. J.*, vol. 18, no. 14, pp. 5805–5811, 2018.

153. A. Leal-Junior, A. Frizera, C. A. F. Marques, and M. J. Pontes, "Mechanical properties characterization of polymethyl methacrylate polymer optical fibers after thermal and chemical treatments," *Opt. Fiber Technol.*, vol. 43, pp. 106–111, 2018.

154. S. Acheroy et al., "Thermal effects on the photoelastic coefficient of polymer optical fibers," *Opt. Lett.*, vol. 41, no. 11, pp. 2517–2520, 2016.

155. A. Theodosiou, A. Lacraz, A. Stassis, M. Komodromos, and K. Kalli, "Plane-by-Plane femtosecond laser inscription method for single-peak Bragg gratings in multimode CYTOP polymer optical fibre," *J. Light. Technol.*, vol. 35, no. 24, pp. 5404–5410, 2017.

156. K. Bhowmik, G. Rajan, E. Ambikairajah, and G.-D. Peng, "Hydrostatic pressure sensitivity of standard polymer fibre Bragg gratings and etched polymer fibre Bragg gratings," in *23rd International Conference on Optical Fibre Sensors*, J. M. López-Higuera et al., Eds. Bellingham: SPIE, 2014, vol. 9157, p. 91573G.

5 POFBG Sensor Applications

5.1 INTRODUCTION

The ability to measure strain with a fiber Bragg grating (FBG) can indirectly be used to quantify other parameters. This becomes possible by embedding FBGs in special arrangements, materials or a combination of both, allowing the conversion of the external stimulus (that can be either mechanical, chemical, or thermal) in strain into the FBG. In this way, fiber optic sensors have been developed within several areas such as medical health care, heritage conservation, civil engineering, among others, allowing to create multifunctional and smart devices. The inherent properties of polymer optical fibers (POFs) such as high elasticity, low Young's modulus, high thermal sensitivity, biological compatibility, etc., combined with the advantages of FBGs, makes this combination widely desired. Thus, a variety of works have been developed exploring those properties. Despite the highly interesting multiparameter sensing capabilities of polymer optical fiber Bragg gratings (POFBGs), the measurement of a single parameter is problematic due to the cross-sensitivity issues. However, this has been mitigated through the implementation of different fiber optic technologies and detection schemes.

This chapter gives a brief review of some of the POFBG applications, as well as the strategies implemented to measure specific parameters. Nevertheless, the multiparameter capability of POFBGs will also be under focus, showing different methodologies used so far for the discrimination between variables.

5.2 POFBG BEND SENSOR

The capability of an FBG to respond under compression and elongation brings the opportunity to develop different fiber optic sensors based on these characteristics. One particular example is bend-monitoring applications.

When an optical fiber fixed to two translational stages and with an initial non-deformed length L_0 (see Figure 5.1a) is bent with an angle θ by moving one of the stages as shown in Figure 5.1b, different fiber portions will be under tension and others in compression.

The arc length at the center of the fiber, in the bent condition shown in Figure 5.1b, is defined as $L_0 = R\theta$, where R is the radius of curvature. While θ is fixed, R depends on the distance from the central fiber axis to any other radius in the fiber cross-section, which we define here as r, (see the right-hand side of

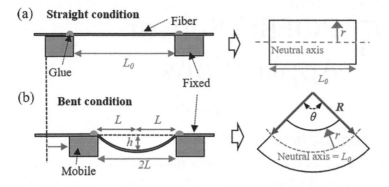

FIGURE 5.1 Schematic of an optical fiber, when it is in a straight condition (a), and when it is bent, by changing the position of the mobile stage to the right (b). The right-hand side of the image provides a close view of the fiber under the straight and bend conditions.

Figure 5.1b). Therefore, the equation that satisfies the fiber length at any curvature radius may be defined as:

$$L = (R - r)\theta \tag{5.1}$$

From Equation 5.1, it is possible to observe that the outermost region is the most stretched region, while the innermost region has the highest compression. Considering conventional fibers, where the fiber core is found in the central region, we have $r = 0$. Therefore, the length at the central region is the same for the straight and for the bent conditions and is thus defined as the neutral axis.

Considering the description of strain along the length on an optical fiber given in Equation 1.14, it is possible to define the strain on a bent condition as:

$$\varepsilon = \frac{L - L_0}{L_0} = \frac{(R - r)\theta - R\theta}{R\theta} = -\frac{r}{R} \tag{5.2}$$

From Equation 5.2, it can be seen that the strain in the fiber core is equal to zero when the fiber length portion is located at the neutral axis ($r = 0$). However, if the core is delocalized from the neutral axis, then the strain in the core can reach higher values. Equation 5.2 also shows that when $R = \infty$ (straight condition), $\varepsilon = 0$ and that decreasing R leads to an increase of the magnitude of ε. Furthermore, the sign of ε will indicate if the fiber length is under compression (negative) or in extension (positive).

After combining Equation 4.6 with Equation 5.2, one can write the Bragg wavelength shift as a function of the bending radius as:

$$\Delta\lambda_C = \lambda_{Bragg}\left(1 - p_e\right)\left(-\frac{r}{R}\right) \tag{5.3}$$

which can be simplified to:

$$\Delta\lambda_C = -\lambda_{Bragg}\left(1 - p_e\right)rC \tag{5.4}$$

where C is the bending curvature given by:

$$C = \frac{1}{R} = \frac{2h}{h^2 + L^2} \tag{5.5}$$

where h is the displacement from the straight position and L is the length of the bent section (see left-hand side of Figure 5.1b). The bending curvature sensitivity is thus expressed by:

$$S_C = -\lambda_{\text{Bragg}}(1 - p_e)r \tag{5.6}$$

The already described advantages of POFs, together with the easiness in offsetting the fiber core from the central region through the fabrication of fiber preforms with simple technologies such as drilling or extrusion, makes them attractive for bend applications. Different authors have been using POF bending sensors in different configurations, such as the use of eccentric core fibers [1], D-shaped POFs through laser micromachining [2] and standard silica and polymer fibers attached to a brass beam [3]. Those configurations may be seen in Figure 5.2a–c, respectively.

The first work reporting the use of a POFBG bend sensor was in 2010 [1,4], for a 230 μm diameter single-mode (SM) step-index POF (SI-POF), in which the core had an offset of 24 μm from its central region. The authors reported orientation dependence in a wide bend curvature range of ±22.7 m^{-1} and with high bend sensitivities of 63 and −56 pm/m^{-1} for the core in tension (when the fiber core is upwards and the

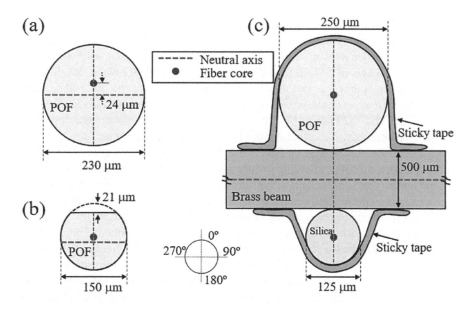

FIGURE 5.2 Different approaches used for bend sensors based on POFBGs. (a) POF with eccentric core [1], (b) surface ablated POF [2], (c) polymer and silica fibers attached to a brass beam with sticky tape [3].

curvature direction is downwards) and compression (when the fiber core is upwards and the curvature direction is also upwards), respectively (see Figure 5.2a).

The possibility of micromachining polymer materials through laser ablation has also been investigated in Ref. [5] in order to provide bend sensitive capabilities [2]. The authors focused the beam of a 248 nm UV laser onto the surface of a 150 μm SM SI-POF. The focused beam was spatially filtered through a square mask (200 × 200 mm) and then scanned along the length of the POF with a fluence of 1.8 J/cm². This allowed to remove part of the fiber cladding, creating a D-shaped fiber (see Figure 5.2b) with an etching depth of ~21 μm along a 10 mm longitudinal length. An FBG was then imprinted on the etched region with a 325 nm UV laser at the 1,550 nm region. The bending characterization was performed in a range of 0–20 m^{-1}, revealing a bend sensitivity of −28 pm/m^{-1} (considering that the D-shape is facing upwards and the curvature tests made downwards, leading to a compression of the Bragg planes).

Recently, the use of a hybrid sensor composed of an in-series silica and polymer FBGs has been reported to simultaneously measure bend and temperature [3,6]. The authors were inspired by the work developed in Ref. [7] where two silica FBGs are attached on the opposite surfaces of an elastic beam for a temperature compensated bend detection. In the work reported in Refs. [3,6], the authors used the same approach, but one of the silica FBGs was replaced by an FBG written in a polymethylmethacrylate (PMMA) SM SI-POF with 250 μm cladding diameter. The fibers were glued in two distinct points along the length of a brass beam with 500 μm diameter. The POF fiber was placed on top of the beam while the silica fiber on the bottom, being both of them fixed with sticky tape (see Figure 5.2c). Such fiber configuration permits to have the fiber cores shifted from the neutral axis, with distances of 375 and 313 μm, for the polymer and silica optical fibers, respectively. According to Equation 5.4, this leads to a higher Bragg wavelength shift for the POFBG, either because of the higher distance of the core from the neutral axis or due to its lower photoelastic coefficient. Curvature characterizations made from −4.5 to 4.5 m^{-1} revealed sensitivities of 533 and −695 pm/m^{-1} for the POFBG when the direction of the curvature is upwards and downwards, respectively. On the other hand, bend sensitivities with almost half of those values were achieved for the silica FBG, which was about 295 and −329 pm/m^{-1} for the curvature tests in the upward and downward direction, respectively. Nevertheless, the proposed fiber optic sensors are also temperature sensitive and because of that, the authors decided to perform a temperature characterization of FBGs, allowing to construct a 2 × 2 matrix containing the curvature and temperature coefficients of both silica and polymer fibers. This matrix could be used for the simultaneous measurement of curvature and temperature (further details on this topic may be seen on Subsection 5.8.1.1). The temperature characterization revealed sensitivities of −147 and 18 pm/°C, for the polymer and silica FBGs. While the POFBG had a similar temperature sensitivity of a freely POFBG, the same did not occur for the silica FBG which showed an increased sensitivity related to its raw value (~10 pm/°C). According to the authors, the reason for such behavior was a consequence of the thermal expansion of the brass beam, which was lower and higher than the ones of the polymer and silica fibers, respectively. The construction of the matrix equation allowed the authors to report resolutions in temperature of 0.012°C, and curvature of 0.008 m^{-1}, considering a detection system of 2.7 pm [3].

5.3 POFBG TRANSVERSE FORCE SENSING

Birefringence in optical fibers occurs due to the presence of asymmetries in its cross section. When optical fibers are subjected to side irradiation for the FBG inscription, a small quantity of photo-induced birefringence appears, leading to polarization effects within the grating [8]. This photo-induced birefringence arises from two main origins which are the polarization of the writing beam and the asymmetric index change in the transverse plane [9]. The combined photo-induced birefringence with the intrinsic fiber birefringence causes the orthogonally polarized modes to experience different couplings through the grating, resulting in an optical signal that is the combination of the two overlapping spectra. However, for this amount of birefringence, it is difficult to distinguish between both, because of the limited resolution of the optical spectrum analyzers, yet it can be clearly seen through polarization-dependent loss (PDL) measurements. PDL can be defined as the maximum change in the transmitted power for all possible states of polarization and is defined as [10]:

$$\text{PDL[dB]} = 10\log_{10} T_{max} - 10\log_{10} T_{min} = 10\log_{10}\left(\frac{T_{max}}{T_{min}}\right) \qquad (5.7)$$

where T_{max} and T_{min}, denote the maximum and minimum power reflected or transmitted through the device. Besides intrinsic birefringence and photo-induced birefringence, external parameters such as twist, bend, transverse load, etc., can also induce small additional birefringence and thus, PDL measurements could be useful for the measurement of these quantities [10].

The photo-induced birefringence in silica optical fibers is well documented [9]; however, for POFBGs, the first study regarding this effect was only studied in 2014 by Hu et al. [11]. Their work reported the measurement of PDL of a 6 mm grating photo-written in a photosensitive SM SI-POF through the phase mask technique and using a 30 mW continuous-wave (CW) Helium-Cadmium (HeCd) laser. The result was an estimated photo-induced birefringence of 7×10^{-6}. Similar results were obtained one year later by the same authors for other types of POFs [12]. To take advantage of the possibility to create a transverse force sensor with temperature insensitiveness as explored in a prior work developed for silica FBGs [13], the authors decided to use the same procedure with POFBGs. An experimental setup was created in which two POFs (one containing an FBG, while another was used as a support) were placed side-by-side in between two 20 mm metal plates, as illustrated in Figure 5.3.

The experiment was made by applying the load on the top plate and it was focused on small transverse force values, for which the transmitted amplitude spectra cannot be reliably used to evaluate the birefringence value, due to the strong overlap between the orthogonally-polarized spectra [12]. The experiment was made for different fiber orientations, in order to check the influence of the transverse force with the orientation of the fiber, being the 0° orientation corresponding to the plane perpendicular to the laser incidence used in the photo-inscription process (see the right-hand side of Figure 5.3). The use of the PDL curve for the transverse force experiments revealed an angular dependence with the fiber orientation, showing positive and negative sensitivity values with a periodicity of 180°. The maximum values were obtained for the 90° and 270°

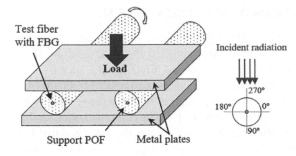

FIGURE 5.3 Schematic of the experimental setup used for the transverse force test used in Ref. [12].

orientations, reaching values of 1.75 and 2.57 dB/N, respectively [12]. The 270° fiber orientation was coincident with that of the laser incidence. The authors also observed that for force values above a few Newtons, there was always a positive slope in the PDL for any orientation. This result was attributed to the higher mechanical-induced birefringence when compared with the photo-induced one. However, in those cases the use of PDL measurements would no longer be an advantage and more suited methods based on the wavelength splitting between orthogonal polarization modes would be more attractive. Furthermore, the authors decided to characterize the fiber sensor to temperature, revealing an almost negligible influence on the PDL [12].

5.4 DYNAMIC POFBG SENSORS

Due to the low Young's modulus, POFs are well suited for dynamic applications, such as the ones used for accelerometers, microphones, and sensors for fast vibration detection. However, POFs are known by their viscoelastic nature (see Subsection 4.2.2), which can pose limits on the final sensing application. In view of that, Stefani and co-workers decided to test POFBGs under dynamic excitations [14]. Their work comprised both PMMA and TOPAS® fibers and it was shown that under low elongations (0.28%) and low frequencies (between 10 and 100 kHz) the fiber sensors showed a frequency-independent flat response, indicating an elastic-like behavior and a low viscosity regime. The sensors response under sinusoidal excitations with frequencies of 1 and 10 Hz and amplitudes of 0.28% and 1.63% was also acquired. Results revealed that no important phase lag between the excitation and sensors response existed. A weak viscoelastic behavior was also observed, leading the authors to conclude that the sensors could be used for high-frequency applications but in low deformation regimes. Those observations were advantageous for the demonstration of a highly sensitive POFBG accelerometer capable to respond to accelerations up to 15 g with a flat frequency response to over 1 kHz [15]. The accelerometer was based on a transducer with a fork shape, with the POFBG bonded in each arm. The characterization was performed by placing the POFBG accelerometer on top of a current driven shaker as shown in Figure 5.4.

The POFBG accelerometer working principle was based on the conversion of acceleration into strain through a mechanical transducer that causes an increase

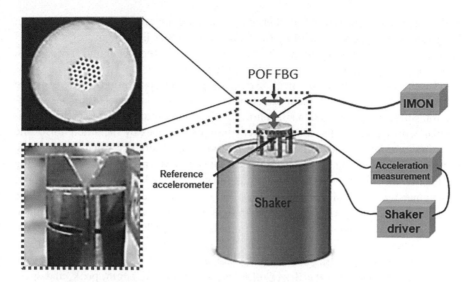

FIGURE 5.4 Schematic of the accelerometer characterization setup, consisting of a POFBG (top inset), attached to a fork shape transducer (bottom inset), placed on a current driven shaker. (Obtained from A. Stefani, et al., "High sensitivity polymer optical fiber-Bragg-grating-based accelerometer," *IEEE Photonics Technol. Lett.*, vol. 24, no. 9, pp. 763–765, 2012, with permission from IEEE© (2012).)

in the separation between the arms of the fork where the POFBG is bonded. The POFBG characterization revealed a linear dependency of the Bragg wavelength shift with acceleration. A comparison of the POFBG accelerometer with the ones based on silica optical fibers revealed a sensitivity improvement by a factor of almost four [15].

Using the same working principle of that presented for the accelerometer, in which the acceleration is transduced into strain in a POFBG by means of a mechanical fork, Stefani et al. have also shown the first results of a POFBG microphone [16]. They placed a membrane in contact with the mechanical fork, converting the sound pressure into elongation on the POFBG and then into a wavelength shift. As a proof of concept, the authors characterized the POFBG microphone to a sound frequency range between 50 and 2 kHz (covering part of the normal voice range). The sensitivity showed a flat response up to 300 Hz with a strong resonance appearing at ~500 Hz. These observations led the authors to conclude that the sensitivity would be enough for the detection of sound at normal speech level but limited by the low narrowband noise level (0.003 pm) [16]. Another interesting work on this topic was performed by Bundalo [17] who explored two different concepts: the use of just a POFBG acting as a microphone and a POFBG attached perpendicularly by its tip to a membrane. Regarding the latter, the author glued the tip of the POFBG to a half-inch thin membrane of a microphone (Brüel & Kjær) and secured the other end at about 5 cm. Pre-tension of the fiber and membrane were adjusted to achieve maximum output response. Results showed that the displacement produced by the membrane was higher than the one measured by the POFBG. Furthermore, the results did not show a uniform and satisfactory frequency response, which was attributed to the internal resonances of the fiber, e.g., viscous dampening, Poisson's ratio, etc. Despite that, studies performed by the same author regarding the use of a POFBG alone (etched one) showed the possibility to considerably recover sound frequencies, which could be used for certain acoustic applications.

The use of acoustic waves in POFBGs for sensing applications has also been investigated for the measurement of viscosity in glucose solutions [18]. A piezoelectric transducer that is fixed to a silica horn as shown in Figure 4.15a, launches longitudinal acoustic waves that are converted into flexural ones by the silica horn. The acoustic waves are transferred to the POFBG which is secured between the tip of the horn and the bottom region of the container. The flexural waves that propagate along the fiber create micro-bends that modulate and perturb the FBG reflectivity. The use of polymer instead of silica FBGs allows larger strain levels for the POFBG case, allowing to tune the grating reflection profile more easily due to its lower mechanical stiffness [19]. The amplitude of displacement is directly related to the fluid's viscosity where the fiber is immersed and thus, an increase in viscosity, forces a decrease of the displacement, leading to a decrease on the perturbation effects. Based on that, the authors characterized the Bragg reflection spectra (wavelength, peak, and spectral bandwidth) and reported sensitivities of -94.42, 2.50 and $-4.95\%/mPa\cdot s$ with resolutions of 0.06, 0.3, and 0.3 mPa·s, respectively [18].

The use of ultrasounds in the medical field have a wide spectrum of applications that ranges from functional and anatomical diagnoses to therapeutic treatments. Studies performed by Gallego and Lamela on ultrasonic sensors based on POFs revealed sensitivities nine times higher (over 1–10 MHz range) compared to their silica counterpart [20]. The result was related with the much lower acoustic impedance and Young's modulus of the POF. Therefore, works regarding the use of POFBG ultrasound sensors for optoacoustic imaging have already been subject of study [21–24]. Tests have been performed for FBGs written either in PMMA as well as TOPAS® fibers, showing similar sensitivities, broadband detection, and lateral directivity. However, PMMA fiber presented higher response time due to the time-dependent process associated with the water absorption.

5.5   POFBG HYDROSTATIC PRESSURE AND LIQUID-LEVEL SENSORS

The capability of a POFBG to respond to a hydrostatic pressure change depends on the amount of strain and refractive index induced on the POFBG, due to fiber compression. Due to the lower Young's modulus of POFs, hydrostatic pressure sensitivities can be several times higher than the ones reported for their silica counterparts, that is about 3 to 4 pm/MPa (see Table 4.5). In addition, by embedding the fibers in special materials, configurations or a combination of both, it is possible to deform the fiber more efficiently, allowing to have higher strain sensitivities. One of those fiber arrangements consists of attaching FBGs to diaphragms.

Diaphragms are deflected when subjected to an external force. In such conditions, the maximum deformation that can occur in the middle region of a circular diaphragm, within the elastic condition and considering clamped edges, may be expressed through the following equation [25]:

$$\delta = \frac{3P\left(1 - v_d^2\right)r_d^4}{16t^3 E_d} \tag{5.8}$$

where P is the applied pressure while v_d, r_d, t, and E_d, are the Poisson's ratio, radius, thickness, and Young's modulus of the diaphragm, respectively. The radial strain at the center of the diaphragm may also be expressed as [26]:

$$\varepsilon_r = \frac{3P\left(1 - v_d^2\right)r_d^2}{8E_d t^2} \tag{5.9}$$

From this equation, it is possible to understand that polymers and rubbers are good material choices due to their low Young's modulus, allowing high radial strains. It is also worth to mention that the fabrication process can be tailored to enhance the radial strain through the production of large radius diaphragms with thin thickness.

5.5.1 POFBG Pressure Transducers

The mechanical deformation of a diaphragm subjected to a hydrostatic pressure can be transferred to an FBG through different configurations. Those may be done either by attaching the FBG onto the diaphragm (embedded [27], or surface-bonded [26]) or by holding a pre-strained FBG between the inner part of the diaphragm and the opposite wall of the box housing [28]. Both configurations are shown in Figure 5.5a and b, respectively.

When the devices shown in Figure 5.5a and b are subjected to hydrostatic pressure, the radial strain on the diaphragm will be transferred to the optical fiber, being the amount of strain transfer dependent on the difference of the mechanical properties of both materials. Thus, considering the ideal case where diaphragms with low Young's modulus are used (e.g., plastics and rubbers), POFs are better candidates to encode the pressure-strain information rather than silica optical fibers. This occurs because the stiff nature of silica fibers could locally reinforce the material.

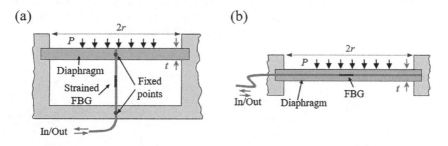

FIGURE 5.5 Schemes used for the conversion of hydrostatic pressure into strain on the FBG by the attachment of a pre-strained FBG between the center of the diaphragm and one of the walls of the box housing (a) [28] and by embedding the FBG along with the diaphragm (b) [27].

Considering that the strain on the diaphragm is completely transferred to the FBG (assuming similar mechanical properties), the FBG pressure wavelength response may be written by substituting Equation 5.9 into Equation 4.6, producing:

$$\Delta\lambda = \frac{\lambda_{\text{Bragg}}\left(1-p_e\right)3P\left(1-v_d^2\right)r_d^2}{8t^2E_d} \tag{5.10}$$

Using the configuration shown in Figure 5.5a, Rajan et al. [28] demonstrated that a 7.5 cm pre-strained POFBG attached to a vinyl diaphragm with a thickness of 60 µm and radius of 4 mm was an interesting scheme to measure hydrostatic pressure, obtaining a sensitivity of 1,320 nm/MPa. The result was due to the pressure-induced deformation on the diaphragm that reduced the amount of strain on the pre-strained POFBG, leading to a blue wavelength shift. Additionally, an epoxy resin with similar mechanical characteristics of the diaphragm was used to allow a complete strain transfer from the vinyl diaphragm to the POFBG. Furthermore, the fiber was also etched to a diameter of 91 µm, allowing to reduce its Young's modulus and enhancing the strain transfer. Later on, in another publication [29], the same authors demonstrated an alternative pressure transducer scheme (Figure 5.6a), intended for blood pressure measurements. The sensor was composed of a perforated aluminum cylinder with a notch at the middle of the longitudinal length and a 150 µm etched POFBG going through the aluminum central hole and glued in both ends. A silicone tube was used around the aluminum cylinder to act as a diaphragm, transferring the hydrostatic pressure to the POFBG underneath. The sensor prototype may be seen in Figure 5.6b.

When the sensor was subjected to hydrostatic pressure, the silicone diaphragm deflects, introducing compressive strain to the POFBG. The characterization of the fiber sensor for a pressure range from 0 to 0.05 MPa, showed a sensitivity response of 1.2 nm/MPa [29].

A different type of pressure transducer based on the use of gratings in POF has been proposed recently by Bundalo et al. [30] using a pod-like structure, in which a pre-strained microstructured polymer optical fiber long-period grating (mPOF-LPG) is attached to the ends of the structure. The principle of operation was based on the conversion of pressure into longitudinal strain on the LPG, allowing to achieve a pressure sensitivity of 105 nm/MPa. This result was higher than the one achieved for silica fiber Bragg gratings reported in other works (~9 nm/MPa). According to the authors, a pressure-to-strain conversion of 0.95% was obtained while the silica

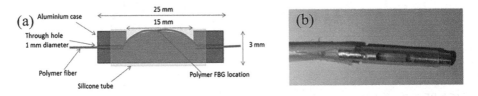

FIGURE 5.6 Schematic (a) and photograph (b) of the pressure transducer. (Reprinted with permission from G. Rajan, et al., "Etched polymer fibre Bragg gratings and their biomedical sensing applications," *Sensors*, vol. 17, no. 10, p. 2336, 2017.)

FBG equivalent was 0.075%. The difference in the results was mainly attributed to Young's modulus difference which benefits the POF, due to the better strain transfer. Giving the strain range covered in this work, which was up to 15 kPa, the authors concluded that the proposed sensor was well suited for in-vivo applications, showing sensitivities one order of magnitude higher than the ones reported in literature [30].

5.5.2 POFBG Liquid Level Transducer

If the hydrostatic pressure is induced through a column of liquid with height h_l, Equation 5.10 becomes:

$$\Delta\lambda_{\text{Bragg}} = \frac{\lambda_{\text{Bragg}}\left(1-p_e\right)3\rho g h_l\left(1-v_d^2\right)r_d^2}{8t^2 E_d} \tag{5.11}$$

where ρ is the liquid density and g is the gravitational acceleration. Changes in liquid level height acting on one side of a diaphragm cause its deflection. By embedding an FBG in a silicone rubber diaphragm, as schematized in Figure 5.5b, a strain transfer occurs between both, leading to a Bragg wavelength shift. This approach has been used recently in Ref. [27], where a silica and polymer FBGs have been embedded in a silicon rubber diaphragm with thickness of ~1 mm and radius of 9.5 mm. Results showed that a liquid level of 75 cm induced a Bragg wavelength shift of 4.3 nm for the POFBG, corresponding to a sensitivity of 57.2 pm/cm, which was five times higher than the one obtained for the same configuration using a silica fiber Bragg grating [27].

5.6 POFBG pH SENSOR

pH measurement is of great importance in many fields, such as biological, ecological, pollution monitoring, environmental, among others. The inherent advantages of optical fiber sensors compared to electric sensors led to the development of different fiber optic pH sensors using principles such as absorbance [31] and photoluminescence effects, especially fluorescence [32]. However, the disadvantages of these methods are related to the light intensity fluctuations, temperature, and dependency on the concentration of indicators.

Fiber Bragg gratings are inherently sensitive to temperature and mechanical deformation. In view of that, coating or packaging FBG in responsive materials can extend their range of sensing capabilities. Hydrogels are a class of hydrophilic polymers that can absorb water and swell or expel water and contract in the presence of an external stimulus. The volumetric change generates a mechanical response in the form of stress that can be readily assessed by combining the hydrogel with a stress-sensitive device [33]. FBGs are one of those devices that can easily measure the amount of strain generated by the polymer absorption or desorption upon an external stimulus. Regarding pH detection, protonation and deprotonation of acidic or basic pendant groups on the polymer material, cause a pH-dependent osmotic pressure difference, which leads to the swelling or shrinkage of the polymer relative to the external conditions [34]. Polymer optical fibers have been reported to have stress sensitivities 28 times higher than the ones reported for silica optical fibers [35]. Hence, the attractiveness of the combination of the swelling properties provided by some special hydrogels and the high-stress sensitivities achieved by POFBGs is highly interesting. This lead recently the Cheng et al. [36] and also Janting et al. [37] to report pH sensors based on the use of a POFBGs encapsulated in a hydrogel material. The configurations used in both works may be seen in Figure 5.7a and b, respectively.

Regarding the work developed by Cheng et al. [36], the authors wrote an FBG in a 125 µm SI-PMMA POF with a core composed of 5% benzyl methacrylate (BzMA) and 1 wt% benzildimethylketal (BDK). The POFBG was encapsulated with poly (ethylene glycol) diacrylate (PEGDA), in molds of 15-mm length and cross-sectional dimensions of 1×1 mm, 0.75×0.75 mm, and 0.5×0.5 mm. Characterizations were made for different aqueous solutions of hydrochloric acid with pH values varying between 6.5 and 0. Results revealed that the Bragg wavelength was unaltered for pH values between 3 and 0 but presented a red wavelength shift when the pH is varied from 6.5 to 3 (see Figure 5.8a). The result was explained as a consequence of the hydrogel swelling promoted by the combination of hydrogen ions with the oxygen atoms in the PEGDA [36], leading to a longitudinal strain on the POFBG. Concerning the influence of the cross-sectional

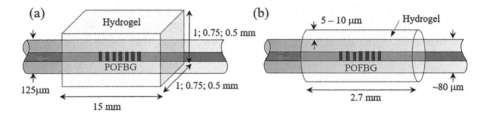

FIGURE 5.7 POFBGs encapsulated in different hydrogel configurations, used for the measurement of pH for the works reported in Ref. [36] (a) and Ref. [37] (b).

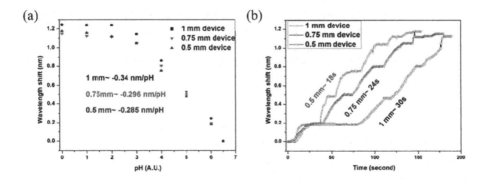

FIGURE 5.8 Wavelength (a) and time response (b) of POFBGs encapsulated with different thickness of PEGDA. (Adapted from X. Cheng, et al., "All-polymer fiber-optic pH sensor," *Opt. Express*, vol. 26, no. 11, pp. 14610–14616, 2018, with permission from the Optical Society of America.)

area on the sensor response, it was observed a sensitivity increase from −285 to −340 pm/pH (see Figure 5.8a) and response time decrease from 18 to 30 seconds (see Figure 5.8b) for the sensors with cross-sectional dimensions of 0.5×0.5 mm and 1×1 mm, respectively.

To further complete their work, and motivated by the stress sensitivity improvement that occurs in etched POFBGs [38], the authors decide to study the influence of the fiber diameter on the sensor pH sensitivity. For that, they used fibers with diameters of 125 μm (raw fiber) and etched fibers with diameters of 102 and 92 μm, each one containing a POFBG covered with PEGDA (cross sections of 1×1 mm). Results showed a sensitivity increase for thinner fibers, reaching values of −340, −380, and −410 pm/pH, for fiber diameters of 125, 102, and 92 μm, respectively [36]. Temperature characterizations were also performed, revealing a positive wavelength shift with increasing temperature, achieving a sensitivity of 10 pm/°C, as a consequence of the large thermal expansion coefficient of the PEGDA.

Regarding the POFBG-based pH sensor developed by Janting et al. [37], schematized in Figure 5.7b, it was used an FBG in an etched PMMA-mPOF (~90 μm) coated with hydrogel. The hydrogel was synthetized using 76 wt% (hydroxyethyl) methacrylate (HEMA) as the backbone of the polymer; 3 wt% ethylene glycol dimethacrylate (EGDMA), acting as a cross-linker between the HEMA chains; 20 wt% methacrylic acid (MAA), used to make the polymer pH sensitive (swelling), and finally 1 wt% 2,2-dimethoxy-2-phenylacetophenone (DMPA) to act as the photoinitiator. The etched mPOF containing the POFBG was then coated with several layers of the hydrogel, giving a final sensing region length of 2.7 mm and thickness between 5 and 10 μm (see schematic in Figure 5.7b). Characterizations performed to the sensor, for $3 <$ pH < 9, revealed an almost insensitive response for pH values below 5, positive wavelength response for $5 <$ pH < 8 and negative response for pH > 8. The sensitivity obtained for $5 <$ pH < 7 was about 73 pm/pH, which was much lower and opposite to the one reported by Cheng et al. [36] (sensitivity of −410 pm/pH for a POFBG with similar fiber dimensions and coated with 1×1 mm PEGDA, for the same pH range). The explanation for the red wavelength shift was described as the increase of the strain on the POFBG as a consequence of the hydrogel swelling with the pH increase. Concerning the blue wavelength shift experienced for pH > 8, the authors mentioned that the increase of the solution ionic strength caused de-swelling of the hydrogel [37]. For the sensor response time, it was reported values below 4.5 minutes for the pH increase/decrease, which was higher than the 30 seconds reported in Ref. [36]. Temperature characterizations of the sensor were also performed in a test tube filled with a buffer solution (pH = 7). A sensitivity of ~−31.2 pm/°C was obtained, which was similar to the etched and uncoated POFBG concluding that the coating has an insignificant effect on the temperature sensor response. Because of that, temperature cross-sensitivity compensation could be easily solved by placing another POFBG with a different resonance Bragg wavelength in the vicinity of the hydrogel POFBG pH sensor.

To conclude the analysis taken for the proposed POFBG pH sensors reported in Refs. [36] and [37], it is important to note that both POFs were composed of PMMA, which is also a hygroscopic material [39]. Consequently, one may assume that for long-term measurements this property would also need to be taken into account. Therefore, humidity-insensitive POFs are an opportunity for this type of sensors [40].

5.7 POFBGS IN HEALTH CARE

The flexibility, non-brittle nature and biocompatibility of POFs led different authors to focus their attention to the biomedical field. Thus, different sensors have been demonstrated namely to monitor blood pressure [29], pressure in endoscopic applications [30], optoacoustic endoscopy [21–23], heartbeat monitoring [41], respiratory monitoring [29,41,42], erythrocyte detection [43], foot plantar pressure [44], and also human-robot interaction forces [45].

5.7.1 RESPIRATORY AND HEARTBEAT SIGNALS WITH POFBGS

Smart clothing is recognized as a key technology for healthcare applications. On the other hand, the fibrous nature of optical fibers makes them well suited for integration into textiles. Because of the high elongation capabilities and the non-brittle nature of POFs, they have been pointed as an opportunity for such applications [46]. The first work intended to use POF gratings in biomedical applications was reported in 2008 [42] for an mPOF-LPG stitched onto an elastic textile fabric. The work was intended to monitor the respiratory rate, either at the thorax or abdomen locations [42]. The suitability of the sensor was tested by clamping the textile in both ends of a traction setup and 25 strain cycles were performed for strains between 0% and 1%, simulating the respiratory movement. Results have shown that the LPG resonance dip wavelength could easily follow the imposed strain steps, demonstrating the feasibility of the use of mPOF LPG for monitoring purposes [42].

Recently, the breathing rate monitoring using POFBGs has also been investigated using different approaches, either by packaging etched POFBGs in a nebulizer mask [29] or by attaching a POFBG through a medical tape to the patient chest [41]. The work performed in Ref. [29] was based on etched POFBGs because the lower Young's modulus would allow to obtain higher pressure sensitivities. For that reason, a SI-SM PMMA POFBG was etched to a diameter of 18 μm and packaged into a perforated plastic tube with 3 mm diameter and 60 mm length, being then attached to the nebulizer mask. Results revealed that the Bragg peak can effectively monitor the normal and accelerated breathing of a volunteer. A Bragg wavelength shift of ~150 pm between the inhale and exhale was reported, as shown in Figure 5.9.

In the most recent work on the respiratory movement using POFBGs [41], the authors attached a POFBG to the chest of a volunteer using medical tape and the Bragg wavelength shift was acquired during the time (see Figure 5.10a). The data was then filtered with low- and high-cut frequencies of 0.15 and 0.3 Hz, respectively, allowing to get the respiratory signal with an amplitude of ~80 pm (see Figure 5.10b). By filtering the raw data with 2 and 8 Hz, it was possible to measure heartbeat with an amplitude of 17 pm (see Figure 5.10c). The same experiments were performed with a silica FBG, revealing values thirty times lower than the ones achieved with the POFBG configuration for both respiratory function and heartbeat monitoring [41].

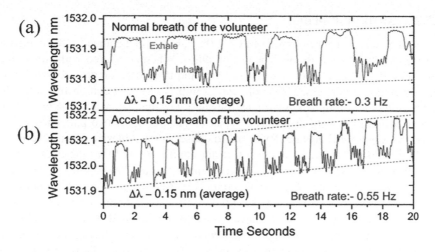

FIGURE 5.9 Normal (a) and accelerated (b) breathing pattern of a volunteer measured by an etched POFBG attached to a nebulizer mask. (Reprinted with permission from G. Rajan, et al., "Etched polymer fibre Bragg gratings and their biomedical sensing applications," *Sensors*, vol. 17, no. 10, p. 2336, 2017.)

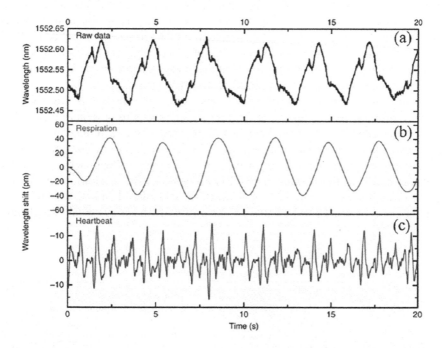

FIGURE 5.10 Waveforms collected from Ref. [41], for the simultaneous measurement of respiration (b) and heartbeat (c), obtained after filtering the raw peak Bragg wavelength (a). (Reprinted from J. Bonefacino, et al., "Ultra-fast polymer optical fibre Bragg grating inscription for medical devices," *Light Sci. Appl.*, vol. 7, no. 3 (2018), 17161, with permission from Springer Nature.)

5.7.2 EMBEDDED POFBGs

Due to the mechanical properties of POFs, they are a suited choice for the detection of strain in materials with low Young's modulus. When compared with silica optical fibers, POFs bring a better strain transfer and therefore, a more accurate strain reading could be performed. One of such examples was first demonstrated for the monitoring of strain in hand-woven textiles [47]. For that, silica and polymer FBGs were attached to a textile using two types of glue, either a two-part epoxy resin as well as a polyvinyl acetate (PVA)-based conservation adhesive. Results revealed that POFs provide a favorable strain transfer coefficient and that the ones bonded with PVA offered less structural reinforcement due to the low Young's modulus of the glue ($E_{PVA} = 0.006 < E_{epoxy} = 1.5\,\text{GPa}$) [47]. Another example that highlights the opportunities of POFBGs for the monitoring of compliant structures has been reported in Ref. [48] targeting the development of flexible and stretchable skins for applications in robotics, healthcare, and structural health monitoring. The comparison between the two FBG responses was made by embedding the fiber sensors in thermally curable polydimethylsiloxane (PDMS) which is a flexible and stretchable material [48]. The embedded FBGs were subjected to a strain test by clamping the flexible skin at both ends and straining it from 0% to 1% (see the setup scheme in Figure 5.11a). Results revealed that despite the similarity on the intrinsic strain sensitivity of the silica and polymer FBGs, when embedded, the POFBG could achieve a strain sensitivity fourty five times higher. Furthermore, a considerable hysteresis was observed for the silica one [48]. The authors also characterized the embedded sensors to pressure. For that, a small weight was placed in different locations along the fiber length (Figure 5.11b). Results revealed that pressure sensing was approximately ten times higher than that of the embedded silica FBG [48].

The use of embedded POFBGs for biomedical applications has already been demonstrated for the monitoring of the foot plantar pressure in dynamic gait [44]. For that, the authors used a 10 mm cork insole grooved with a 2.5 mm deep and 2 mm wide across key areas for foot plantar pressure namely heel, midfoot, metatarsal, and toe. These specific areas were further caved in circular shapes with a diameter of 10 mm and a depth of 5 mm. A POFBG array with FBG separation length preselected to fall within the middle region of those areas was then inserted and filled with an epoxy resin. Each POFBG was able to shift the Bragg peak wavelength

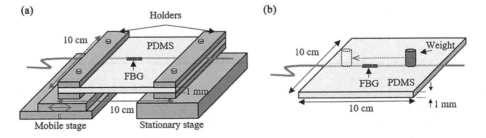

FIGURE 5.11 Schematic of the strain (a) and pressure (b), characterizations, performed for embedded FBGs in a PDMS substrate, accomplished in Ref. [48].

accordingly to the pressure-induced strain and thus, capable to monitor the foot plantar pressure. Results revealed similar performance to the ones fabricated for the same system using silica FBGs [49]. However, the sensitivity was doubled and the benefits of high flexibility and non-brittle nature of POFs is more appealing.

The use of POFBGs has also been under investigation for simultaneous measurement of temperature and thermal expansion effects in a composite material [50]. The composite was fabricated by packaging eight layers of a glass fiber fabric with polyester resin as the matrix material. The POFBG was placed in the middle region of the 2nd layer and pre-strained before hardening of the material. The characterization results revealed that the embedded POFBG temperature sensitivity was similar to that of the raw POFBG. However, an increase of the spectral bandwidth associated with the high transversal thermal expansion coefficient was observed. Therefore, the authors concluded that it was possible to discriminate between temperature and thermal expansion effects on the composite material [50].

POFBGs embedded in plastics fabricated through three-dimensional (3D) printers, such as polylactic acid (PLA) and acrylonitrile butadiene styrene (ABS) [51], as well as flexible thermoplastic polyurethane (TPU) [45] have also been reported in literature for strain and temperature sensing [51] and for force and temperature on the assessment of human-robot interaction forces in a wearable exoskeleton [45]. Regarding the latter, the authors printed in 3D two rigid ABS structures to support the flexible 3D printed TPU material used to embed the POFBG. The structure was attached to the shank region of an exoskeleton for knee rehabilitation and the results showed the feasibility of the proposed approach.

5.7.3 Erythrocyte Detection

The biocompatibility of POFs makes them advantageous for biosensing applications. Nevertheless, POFs can be made of biodegradable materials [52] or can be modified to detect specific biochemical species [53,54]. Furthermore, the advent of POFBGs brought new sensing opportunities due to the easy integration and characterization. On the other hand, graphene has the unique ability to dramatically promote sensitivity in biochemical detection, i.e., both the conductivity and the permittivity of graphene are ultrasensitive to external molecular adsorptions [55].

Based on these opportunities, in 2015, Yao et al. reported the development of a D-shaped POFBG covered with a monolayer of p-doped graphene for highly sensitive erythrocyte detection [43]. For that, the authors inscribed a POFBG at the infrared region in a 3-ring 110 µm PMMA mPOF. The fiber was then side polished close enough to the core region in order to allow the exposure of the evanescent field to the external medium (length of 8 mm and depth of ~50 µm). The authors have then deposit a p-doped graphene monolayer by chemical vapor deposition onto a copper foil that was posteriorly transferred onto a PMMA film used to cover the D-shaped section. The PMMA thin film had ~50 nm width and was used to keep the graphene layer (~0.4 nm) doped and to help cell adsorption. The schematic diagram of the cross-sectional view of the sensor may be seen in Figure 5.12.

The characterization tests were made in a microfluidic system using the D-shaped POFBG with and without the graphene layer. The tests were made for four fresh red blood

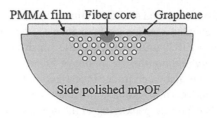

FIGURE 5.12 Schematic diagram of the graphene-based D-shaped POF sensor (cross-sectional view), used in Ref. [43].

cells/physiological-saline solutions with different concentrations, ranging from 0 ppm (pure saline solution) to ~ 10^4 ppm, corresponding to a refractive index range between 1.334 and 1.337. Results showed that the Bragg peak decreased its power and was red-shifted with increasing concentration. Sensitivities of 0.27 dB/kppm and 0.024 pm/ppm were obtained for the D-shaped POFBG, while sensitivity values of 0.15 dB/kppm and 1 pm/ppm were observed for the D-shaped POFBG covered with the PMMA/graphene film [43], presenting a much higher sensitivity for the latter. The result was a consequence of the more pronounced exposure of the evanescent field to the external medium for the configuration containing the PMMA/p-doped graphene film. When exposing the fiber sensing area to the red blood cells, the permittivity of the p-doped graphene is modified due to the cell adsorption. Therefore, according to the authors, the index of the graphene would be modified, since the permittivity is determined by the conductivity. The results have also shown good recoverability, concluding that the proposed sensor could be useful for highly accurate diagnosis of blood-related diseases like nephritis, hemorrhage diagnosis, infection immunity research, and biological safety [43].

5.7.4 POFBG ACOUSTIC SENSORS

Intravascular ultrasound (IVUS) is an imaging method capable to visualize atheromatous plaque in the coronary arteries. An IVUS catheter works by emitting ultrasonic waves and by receiving the backscattered signal from the tissues. IVUS is thus capable to provide tomographic views that allow an accurate determination of location and morphology of vulnerable atherosclerotic plaques (VAP). These are known to be the major contribution to most of the acute cardiovascular events and sudden cardiac deaths. This occurs because the rupture of a VAP releases thrombogenic contents of the plaque into the bloodstream resulting in acute myocardial infarction or in a stroke with a high mortality rate. The susceptibility of rupture of a VAP is associated with the composition of the plaque, the distribution of mechanical stress within it and the presence and extent of associated inflammation [56]. Despite the valuable information given by IVUS, there is limited specificity for different soft tissues types and thus, difficulties to discriminate the plaque composition. Optoacoustic endoscopy can provide spectroscopic information by mapping the optical absorption distribution [56] and thus, it can complement IVUS. Optical fibers are well suited to be incorporated in an IVUS probe since they are miniaturized in size, immune to electromagnetic interference, made of materials compatible with Magnetic Resonance Imaging (MRI) and provide large detection bandwidth compared to piezoelectric transducers (when miniaturization is taken into account). Consequently, many fiber optic sensors have been proposed such as Fabry-Pérot (FP) interferometers, Mach-Zehnder interferometers (MZIs), and FBGs. Concerning the latter, they appear to be a good candidate since they have a natural "lateral view" contrary to FP fiber sensors, as well as to provide a specific sensing region located at the grating structure, contrary to MZI fiber sensors. FBGs in silica fibers were first demonstrated for detecting ultrasounds by Webb et al. in 1996 [57]. The intrinsic acoustic sensitivity for an FBG can be expressed in terms of the variation of the Bragg wavelength induced by the magnitude of the acoustic pressure as described previously by Equation 4.30. Thus, the use of materials with low Young's modulus plays an important contribution to the sensitivity improvement. Nevertheless, the acoustic impedance of the POF is matched to the water with greater efficiency when compared with the silica optical fiber [20]. Based on those characteristics, POFBGs in PMMA [21–24] and TOPAS® [23] were already proposed for ultrasonic endoscopic applications. The principle of operation is based on the use of a fast photodiode for the detection of the reflected power when the FBG is interrogated with a narrow line laser tuned at 3 dB spectral bandwidth. Regarding the work presented in Ref. [23], it was shown that a POFBGs recorded in TOPAS® fiber was better suited for optoacoustic endoscopy than the one in PMMA due to the long time response associated with the water absorption, for the PMMA case. The characterization of both fiber sensors with a 10 MHz burst and a 55 MHz detection bandwidth revealed a noise equivalent power of 1.31 and 1.98 kPa for the 5 mm long PMMA FBG and 2-mm long TOPAS® FBG, respectively. In this way, the intrinsic sensitivity per unit length was similar for both fiber sensors [23]. Furthermore, the authors have also studied the lateral directivity of both POFBG sensors, showing a sensitivity beyond a frequency of 20 MHz and a lateral directivity at 10 MHz of ±2 degrees in the case of PMMA FBG and ±5 degrees for TOPAS® FBG. The differences were mainly due to the FBG lengths [23]. The development of the technology is still ongoing and the incorporation of π-shifted FBGs to achieve a narrower spectrum with higher sensitivity will be probably the next step to improve the results so far achieved.

5.8 SIMULTANEOUS SENSING WITH POFBGs

Fiber Bragg gratings are well known from their multiparameter sensing capabilities. Regarding POFBGs, special care as to be taken into account due to the large cross-sensitivity to temperature and humidity. Along the years, many techniques have been proposed to discriminate or to remove the influence of unwanted parameters. Due to the simplicity, the most adopted method is based on the use of an extra reference grating. Other approaches include the combination of FBGs with other fiber optic technologies or the use of fibers based on different materials and geometries.

5.8.1 SIMULTANEOUS MEASUREMENT OF STRAIN AND TEMPERATURE

FBGs are inherently sensitive to strain and temperature, creating challenges to discriminate between both parameters. A possible solution can be made through the combination of two fiber optic sensor elements with very different responses to strain

$(S_{\varepsilon 1}, S_{\varepsilon 2})$ and temperature (S_{T1}, S_{T2}), placed at the same location on the structure, allowing to get a matrix equation as follows [58]:

$$\begin{pmatrix} \Delta\lambda_1 \\ \Delta\lambda_2 \end{pmatrix} = \begin{pmatrix} S_{\varepsilon 1} & S_{T1} \\ S_{\varepsilon 2} & S_{T2} \end{pmatrix} \begin{pmatrix} \Delta\varepsilon \\ \Delta T \end{pmatrix} \tag{5.12}$$

where $\Delta\lambda_i$ is the wavelength shift of the sensor element referred to as the subscript i when subjected simultaneously to a change in strain ($\Delta\varepsilon$) and temperature (ΔT). Equation 5.12 can then be inverted to yield the strain and temperature on the structure as:

$$\begin{pmatrix} \Delta\varepsilon \\ \Delta T \end{pmatrix} = M^{-1} \begin{pmatrix} \Delta\lambda_1 \\ \Delta\lambda_2 \end{pmatrix} \tag{5.13}$$

where

$$M^{-1} = \frac{1}{|M|} \begin{pmatrix} S_{T2} & -S_{T1} \\ -S_{\varepsilon 2} & S_{\varepsilon 1} \end{pmatrix} \tag{5.14}$$

being $|M| = S_{\varepsilon 1} \cdot S_{T2} - S_{\varepsilon 2} \cdot S_{T1}$ the determinant of the coefficient matrix.

The success of this technique depends on the difference between the ratio of the strain responses of the two sensors and the ratio of their temperature responses, in such a way that the determinant of the matrix is nonzero [58]. The demonstration of this methodology was studied by Xu et al. taking into account the change of the photoelastic and thermo-optic coefficients of the optical fiber material with wavelength. In this way, the possibility to simultaneously measure strain and temperature using two in-series FBGs written in different spectral windows was demonstrated [59]. However, other methodologies have also been explored along the years, using other fiber optic schemes [58].

5.8.1.1   Dual FBG Sensors

In 2003, four years after the first demonstration of a POFBG [60] and right after the demonstration of their high tuning capabilities with strain [61] and temperature [62], Liu et al [63], reported the use of a silica FBG in-series with a PMMA-based POFBG for the discrimination of strain and temperature (Figure 5.13a). Characterizations of the fiber sensor showed sensitivities to strain of $S_{\varepsilon 1} = 1.5$ pm/με, $S_{\varepsilon 2} = 1.2$ pm/με and sensitivities to the temperature of $S_{T1} = -149.0$ pm/°C and $S_{T2} = 10.5$ pm/°C, where the subscript 1 and 2 define the FBG in polymer and silica optical fibers, respectively. Because the tuning range of the POFBG is much larger than that of the silica FBG, the dynamic ranges of ΔT and $\Delta\varepsilon$ are correspondingly much larger than in the case of pure silica gratings.

The change in the mechanical and thermal properties of optical fibers after etching procedures can also be used to achieve different strain and temperature responses compared to that of raw fibers. Because of that, Bhowmik and co-workers [64,65]

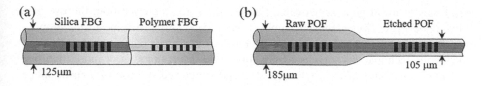

FIGURE 5.13 Configurations used for the simultaneous measurement of strain and temperature, using an in-series (a) silica and polymer FBGs [63] or (b) raw and etched POFBGs [64,65].

reported the use of FBGs at the C-band in a non-etched and etched region of a photosensitive PMMA SI-POF for the simultaneous recovery of strain and temperature (Figure 5.13b). Results revealed that FBGs written in the 185 µm non-etched region of the POF presented sensitivities of $S_{T1} = -92$ pm/°C and $S_{\varepsilon 1} = 1.24$ pm/µε, while FBGs in the etched region (105 µm diameter) revealed higher sensitivities, reaching values of $S_{T2} = -133$ pm/°C and $S_{\varepsilon 2} = 1.65$ pm/µε. Experimental validation of the proposed sensor revealed maximum errors of ±8.42 µε and ±0.39°C.

5.8.1.2 Resin-Based FP Cavity Combined with POFBG

Nowadays optical fiber devices such as sources, detectors couplers, etc., are essentially based on the well-developed silica optical fiber. Hence, the operation of a POF sensor usually needs to have some type of interconnection between both. This has been mainly implemented through the termination of POFs in fiber optic connectors that can be easily coupled to connectorized silica fibers using fiber optic adaptors [66]. However, the production of POFs is still facing some problems such as lack of standardization that ultimately lead to fluctuations on the diameter and on the core concentricity along the fiber length [67]. Nowadays, a "cold splice" using a photopolymerizable resin between the silica and polymer fiber is commonly accepted (see Section 1.4). The process is made by aligning the fiber cores through 3D stages while a drop of a photopolymerizable resin with good transparency, mechanical resistance, and refractive index matching one of the fibers, is used to provide a secure connection between the fibers. Because of the refractive index mismatch between the silica and POFs, any gap between the fibers may cause the formation of a Fabry-Pérot cavity. The formation of such structure will impose a wavelength-dependent intensity modulation of the input light spectrum that is caused by the optical phase difference between the reflected beams on the two fiber interfaces. The period of modulation ($\Delta\lambda$) is a function of the wavelength of the radiation (λ), the refractive index of material within the cavity (n), and distance between the fibers (D_L), and it can be written as:

$$\Delta\lambda = \frac{\lambda^2}{2nD_L} \tag{5.15}$$

Normally, when measuring the reflection signal of a POFBG, the formation of a Fabry-Pérot cavity between the silica and polymer fibers is an undesired effect since it deteriorates the signal. To prevent the formation of such structure, the silica fiber terminal is normally cleaved with an 8° angle to prevent Fresnel reflections. Furthermore, the optical fibers are also placed in contact through their end face terminals to avoid gaps in-between. Despite the signal deterioration, the use of a fiber optic FP cavity in sensing applications is very attractive due to its simplicity, compact size, and sensitivity [68,69]. Furthermore, their combination with other fiber optic technologies such as FBGs allows the discrimination of more than one parameter. Because of that, Oliveira et al. [70] reported the use of a resin splice connectorization with distance of a few microns between a silica lead-in fiber and a POF containing an FBG. This allowed the creation of an FP cavity in-series with a POFBG for the simultaneous measurement of strain and temperature. For the fabrication of the FP cavity, a photopolymerizable resin, namely NOA86H from Norland® Products Inc. was used, which provided a tensile strength similar to that of the mPOF used. Furthermore, the fibers terminals were prepared with FC/PC terminals and then positioned in a 3D axis mechanical stage for alignment through their centers and with a distance of 40 µm inbetween. Then, a drop of resin was deposited onto the fibers gap and cured under UV radiation, allowing to get the FP cavity shown in Figure 5.14a. The POFBG was then written and the prepared dual inline sensor was fixed in a strain characterization stage (total length of 9.7 cm: 5 cm silica fiber and 4.7 cm mPOF). The results concerning the spectral analysis for the temperature and strain characterizations may be seen in Figure 5.14b and c.

After the characterization of the dual inline sensor, the authors report sensitivities of $S_{\varepsilon 1} = 2.4$ pm/µε, $S_{\varepsilon 2} = 1.8$ pm/µε, $S_{T1} = -49.0$ pm/°C and $S_{T2} = 641.4$ pm/°C [70], where the subscript 1 and 2 define the POFBG and resin-based FP cavity, respectively. Regarding the POFBG strain sensitivity, it can be observed that the value was twice the one achieved for a standard POFBG. However, since a piece of the silica lead-in fiber is involved in the strain characterization, an unequal strain distribution

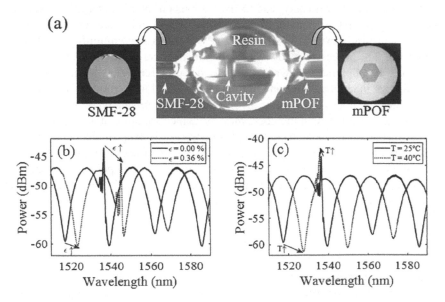

FIGURE 5.14 (a) Resin-based Fabry-Pérot cavity, formed between the silica fiber at the left and by the mPOF at the right. Reflection spectra evolution of the dual inline sensor for the strain (b) and temperature (c), characterizations reported in Ref. [70].

will occur in each fiber segment. Thus, a higher density of strain is reached on the POF section and consequently, a higher POFBG strain sensitivity will be achieved for the POF in this configuration than considering it alone (see Subsection 4.2.3). By using the matrix sensitivity shown in Equation 5.12, the authors estimated a resolution for the strain and temperature of 4.29 µε and 0.02°C, respectively, considering a detection system resolution of 10 pm.

5.8.1.3 Multimode Interference in POF Combined with POFBG

The combination of POFBGs with other fiber optic technologies such as multimode interference (MMI) in POFs has also been proposed for multiparameter-based applications by Oliveira et al. [71]. The authors spliced a 10 cm length of a 73.5 µm diameter unclad ZEONEX® 480R multimode (MM) POF and containing an FBG, between two single-mode silica fibers using a photopolymerizable resin. The schematic of the fiber sensor and its picture transmitting white light may be seen in Figure 5.15a and b, respectively.

After the sensor assembly, the POFBG signal was monitored in reflection while the MMI interference signal in transmission. The corresponding spectra may be seen in Figure 5.16a and b, respectively.

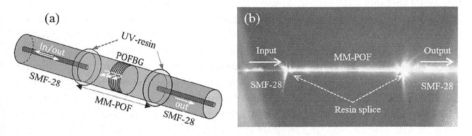

FIGURE 5.15 Schematic of the dual inline sensor used in Ref. [71], composed of a section of a ZEONEX® 480R MM-POF (containing an FBG), in between two single-mode silica fibers (a). Picture of the fiber sensor with white light transmission (b).

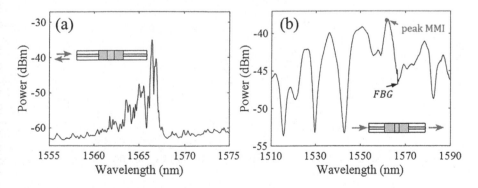

FIGURE 5.16 Reflection (a) and transmission (b) spectra of the inline POFBG and MMI sensor. (Obtained from R. Oliveira, et al., "Multiparameter POF sensing based on multimode interference and fiber Bragg grating," *J. Light. Technol.*, vol. 35, no. 1, pp. 3–9, 2017, with permission from IEEE© (2017).)

The fiber sensor was then characterized to strain and temperature. For that, the splicing regions were secured with additional resin to both ends of a strain characterization setup. The strain range was up to 1.5% while the temperature tests were made from 25°C up to 105°C. The results are shown in Figure 5.17a and b, respectively.

The characterization results revealed strain sensitivities of 1.5 and −3.0 pm/με while the temperature sensitivities were −64.0 and 102.9 pm/°C for the POFBG and MMI structure, respectively. With these results, the authors conclude that a resolution of 40.3 με and 1.1°C could be obtained for a system with 10 pm wavelength resolution and considering the matrix sensitivity shown in Equation 5.12.

Another interesting methodology to discriminate between strain and temperature can be done by using an isolated reference FBG that only monitors the temperature effects [72]. This method has been explored in Ref. [73] using a dual-FBG in a PMMA mPOF. For that, the authors inscribed two in-series and closely separated FBGs at the 850 nm region, using the same phase mask but applying a different amount of strain during the inscription, as explained in Section 3.2.3. The characterization results were performed by subtracting the wavelength shift of the reference grating to the one subjected to strain and temperature. The results demonstrated a temperature compensated POFBG strain sensor with a temperature sensitivity crosstalk of 0.0095 με/°C.

Despite the relatively easy implementation of a dual-FBG sensor, the use of a single FBG for the discrimination of strain and temperature is more desirable. Because of that and taking into account the additional information contained in multi-peak FBG, Qiu et al. reported the use of a few-mode POFBG for the strain and temperature discrimination [74]. For that, the authors wrote an FBG at 1,550 nm region in a PMMA-BDK doped core SI-POF, obtaining multiple peaks between 1,566 and 1,571 nm. Characterization results for all the guided modes revealed different temperature sensitivities, ranging from 98 to 111 pm/°C (i.e., S_{T1} and S_{T2}) for the lower and higher-order modes, respectively, while a sensitivity value of ~1.2 pm/με for all the modes involved (i.e., $S_{\varepsilon 1} \approx S_{\varepsilon 2} = 1.2$ pm/με). These results allowed the construction

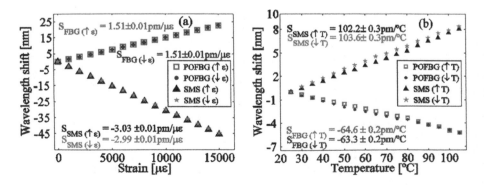

FIGURE 5.17 Wavelength shifts produced by the POFBG and MMI peak power for the strain (a) and temperature tests (b). (Obtained from R. Oliveira, "Multiparameter POF sensing based on multimode interference and fiber Bragg grating," *J. Light. Technol.*, vol. 35, no. 1, pp. 3–9, 2017, with permission from IEEE© (2017).)

of a sensitivity matrix for the simultaneous measurement of strain and temperature (as described in Equation 5.13).

The use of a single Bragg grating with a chirped profile is also feasible to measure strain and temperature as already described for the well-known silica fiber [58]. The mechanism behind the discrimination between both relies on the shift of the Bragg wavelength with strain and temperature and on the broadening of the spectral bandwidth with strain. Taking into account the opportunities of POFs and the possibility to write chirped FBGs in them [75], Min et al. reported the use of a chirped FBG in a three-hole layer BDK doped mPOF at the 850 nm region, allowing simultaneous measurement of strain and temperature [76]. To achieve the chirped fiber Bragg gratings (CFBG), the fiber was tapered in acetone using a controlled translation stage which allowed to obtain the desired taper profile. Then, the CFBG was written on the tapered region. Characterization tests revealed a positive wavelength shift of both central wavelength and an increase of the reflection bandwidth with increasing strain, reaching values of 0.9 and ~0.2 pm/με, respectively. Regarding temperature characterizations, the authors revealed a Bragg wavelength sensitivity of −58 pm/°C without change on the spectral bandwidth, showing that the proposed sensor could be used to effectively recover strain and temperature simultaneously.

5.8.2 Simultaneous Measurement of Humidity and Temperature

PMMA, the most used polymer material, is well known for its high water absorption properties. Therefore, concerns about humidity cross-sensitivity in POFBGs based on this type of material start to appear in 2008 by Harbach [77]. In his work, it was shown that FBGs written in a PMMA SI-POF presented temperature sensitivities of −10, −34, and −138 pm/°C, in dry, wet, and ambient conditions, respectively [77,78].

To suppress the humidity influence in single parameter applications, the research community has implemented a variety of solutions. Among them was the use of novel POF materials such as TOPAS® [40,79,80], ZEONEX® [71,80,81], and CYTOP® [82] which have low water absorption capabilities, leading to the development of humidity insensitive POFBG sensors.

However, the simultaneous measurement of humidity and temperature can be advantageous in applications such as aviation, climate research, semiconductors, and food technology. In view of that, some publications regarding this subject have already been demonstrated using different fiber optic technologies, such as Fabry-Pérot cavities in photonic crystal fibers coated with polyvinyl alcohol [83], dual FP droplet sensor [84], coated and uncoated silica FBGs [85,86], etc. Regarding the latter, the humidity insensitiveness of silica fibers led the authors to coat one of the FBGs with hygroscopic polymers. By doing that, it was possible to measure temperature through a reference FBG and both humidity and temperature through the coated one. The advent of FBGs in humidity-sensitive POFs makes a sensor of this type very appealing. In this way, different fiber optic schemes have been implemented using different methods namely the combination of silica and polymer FBGs [87], the use of FBGs in POFs composed of materials with different hygroscopic properties [88] and also through the use of POFBGs combined with adhesive-based Fabry-Pérot cavities [89,90].

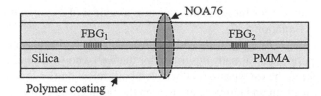

FIGURE 5.18 Schematic of an in-series, silica and polymer FBG, proposed in Ref. [87] for the simultaneous recovery of humidity and temperature.

5.8.2.1 In-Series Silica and Polymer FBGs

Zhang et al. were pioneered in reporting the simultaneous measurement of humidity and temperature using POFBGs [87]. For that, the authors coupled a silica fiber to a POFBG using a UV curable resin as depicted on the schematic presented in Figure 5.18.

Characterization of the fiber optic sensor to temperature showed a sensitivity of −55 and 14 pm/°C for the polymer and silica FBGs, respectively. Regarding the humidity characterization, they obtained sensitivities of 35.2 and 0.3 pm/%RH, respectively. In this particular case, the silica FBG had some humidity sensitivity since the authors recoated the fiber with some type of acrylate after the inscription process (probably to reinforce the FBG section). Considering a matrix sensitivity on the same base like the one shown in Equation 5.12 and a 10 pm system resolution, the system error was about 1.1%RH for the humidity and 0.7°C for the temperature.

5.8.2.2 In-Series POFBGs

The capability to produce POFs based on different polymer materials can bring new features to the fibers. Based on this, Woyessa et al. [88] reported the use of a ZEONEX® based mPOF with a PMMA overcladding, to simultaneously measure humidity and temperature. Depending on the presence or absence of the overcladding region two distinct hygroscopic characteristics were obtained. To create the fiber sensor, the authors etched the PMMA overcladding in one portion of the fiber and then wrote an FBG (FBG₂). Then, the inscription of the second FBG (FBG₁) was followed in the raw fiber region, obtaining two in-series FBGs as represented in Figure 5.19.

The fiber sensor characterizations revealed that FBG$_2$ had a linear behavior with temperature and humidity, obtaining sensitivities of −23.9 pm/°C and 1.4 pm/%RH,

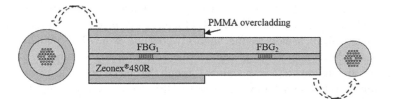

FIGURE 5.19 Schematic of the fiber sensor used in Ref. [88] for the simultaneous measurement of humidity and temperature using FBGs in ZEONEX® 480R mPOF, with and without PMMA overcladding.

respectively. On the other hand, FBG_1 had also a linear behavior with temperature, showing a sensitivity of -15.1 pm/°C, but a quadratic behavior was observed for the humidity characterizations, reporting coefficient values of 6.4 and 0.057 pm/%RH^2. The nonlinear behavior observed for the humidity results is well known from hygroscopic polymers [91]. However, for simplicity reasons, many authors linearize the experimental data to first-order polynomial fits [77]. After resolving a system of equations given by the sensitivity coefficients, the authors concluded that the proposed fiber optic sensor could effectively recover temperature and humidity simultaneously, with resolutions of <0.5°C and <1.2%RH, respectively.

5.8.2.3 Fabry Pérot Cavity Combined with POFBG

Taking into account that the interrogation of POFBGs with standard methods requires the use of a splice between the silica light delivering system and the POF, Oliveira et al. [89] took this property as an advantage for the creation of an FP cavity in between the fibers. On this basis, two inline sensors (FP and POFBG) were obtained, allowing discrimination of two parameters, such as humidity and temperature [89]. For that, they cleaved the fibers perpendicularly to its length, aligned them through their centers and fixed the separation between the two fiber terminals with a length of 30 µm. In this way, a low finesse FP cavity was created. In order to provide a robust fiber splice, they used the photopolymerizable resin NOA86H from Norland® Products Inc., which provided a tensile strength similar to that of the POF. The schematic of the fiber sensor as well as the reflection signal from the inline fiber sensor may be seen in Figure 5.20a and b.

The experimental characterizations were made in reflection by monitoring the wavelength of one of the interference fringes of the FP cavity and also by tracking the Bragg peak wavelength (see Figure 5.20b). Results revealed sensitivity to temperature of 177.5 and -48.6 pm/°C for the FP cavity and POFBG, respectively. Humidity characterizations showed values of 55.1 and 51.1 pm/%RH for the FP cavity and POFBG, respectively [89]. Despite the opportunities of this type of sensor, the creation of fringes with good visibility is difficult to achieve due to the low

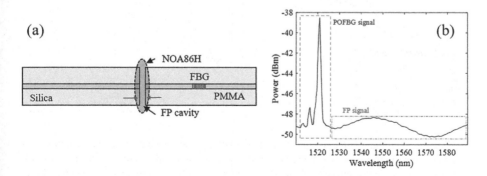

FIGURE 5.20 Schematic of the inline fiber optic sensor used in Ref. [89], for the simultaneous measurement of humidity and temperature using a POFBG combined with resin-based Fabry-Pérot cavity, formed between the fiber terminals of silica and polymer fibers (a). Reflection spectrum of the inline fiber optic sensor (b).

reflectance of the POF interface associated with the polishing process. Regarding Ref. [89], the peak to dip was about 2 dB, while the free spectral range was about 60 nm (see Figure 5.20b). In such cases the detection of the peak or dip wavelength causes enormous errors on the parameter being measured. To suppress this problem, the same authors decided to splice the silica fiber without any gap and with the same UV resin (NOA86H) to a thin flat sides mPOF containing the POFBG [90]. Then, they produced a 30 μm resin-based FP cavity between two silica fibers using NOA78 from Norland® Products Inc. The use of two silica fiber end faces provided better reflectance compared to the silica-POF approach. Nevertheless, the use of NOA78, as well as the use of a thinner POF, was intended to reduce the sensor response time due to the water absorption as demonstrated in Refs. [92] and [68], respectively. The optical outputs from each structure were then combined through a fiber coupler and seen in reflection as represented in Figure 5.21.

After characterizing the fiber sensors to temperature, the authors obtained a temperature sensitivity of −20.6 and 386 pm/°C, for the POFBG and FP cavity, respectively. Concerning the humidity results, they adjusted the experimental data to quadratic functions, obtaining sensitivity coefficients of 52.1 pm/%RH and 0.16 pm/%RH^2 for the POFBG, while values of 333 pm/%RH and 4 pm/RH^2 for the FP cavity. By constructing a system of equations, they were able to report the capability to simultaneous measure humidity and temperature with resolutions below 0.2%RH and 0.2°C, respectively [90], which were below than the ones reported in the literature.

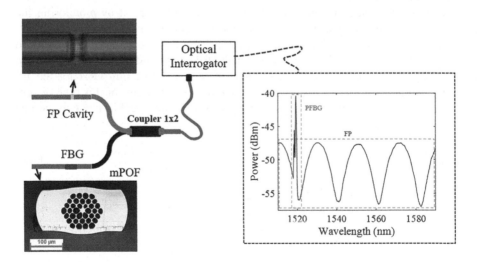

FIGURE 5.21 The left-hand side shows the schematic representation of the measurement scheme used to simultaneous measure humidity and temperature using a resin-based FP cavity formed in between two silica fibers and a POFBG written in a flat sides mPOF. The right-hand side shows correspondent reflection spectrum. (Obtained from R. Oliveira, et al., "Simultaneous detection of humidity and temperature through an adhesive based Fabry – Perot cavity combined with polymer fiber Bragg grating," *Opt. Lasers Eng.*, vol. 114, pp. 37–43, 2019, with permission from Elsevier© 2019)

REFERENCES

1. X. Chen, C. Zhang, D. J. Webb, K. Kalli, and G.-D. Peng, "Highly sensitive bend sensor based on Bragg grating in eccentric core polymer fiber," *IEEE Photonics Technol. Lett.*, vol. 22, no. 11, pp. 850–852, 2010.
2. X. Hu, X. Chen, C. Liu, P. Mégret, and C. Caucheteur, "D-shaped polymer optical fiber Bragg grating for bend sensing," in *Advanced Photonics*, 2015, p. SeS2B.5.
3. B. Yan et al., "Simultaneous vector bend and temperature sensing based on a polymer and silica optical fibre grating pair," *Sensors*, vol. 18, no. 10, p. 3507, 2018.
4. X. Chen, C. Zhang, D. J. Webb, G.-D. Peng, and K. Kalli, "Bragg grating in a polymer optical fibre for strain, bend and temperature sensing," *Meas. Sci. Technol.*, vol. 21, no. 9, p. 094005, 2010.
5. C. Ying-Tung, K. Naessens, B. Roel, L. Yunn-Shiuan, and A. A. Tseng, "Ablation of transparent materials using excimer lasers for photonic applications," *Opt. Rev.*, vol. 12, no. 6, pp. 427–441, 2005.
6. B. Yan et al., "Vector bend sensing based on polymer and silica fiber Bragg gratings," in *Conference on Lasers and Electro-Optics Pacific Rim (CLEO-PR)*, 2017, p. s1422.
7. R. Falciai and C. Trono, "Curved elastic beam with opposed fiber-bragg gratings for measurement of large displacements with temperature compensation," *IEEE Sens. J.*, vol. 5, no. 6, pp. 1310–1314, 2005.
8. N. Belhadj, S. LaRochelle, and K. Dossou, "Form birefringence in UV-exposed photosensitive fibers computed using a higher order finite element method," *Opt. Express*, vol. 12, no. 8, pp. 1720–1726, 2004.
9. T. Erdogan and V. Mizrahi, "Characterization of UV-induced birefringence in photosensitive Ge-doped silica optical fibers," *J. Opt. Soc. Am. B*, vol. 11, no. 10, pp. 2100–2015, 2008.
10. S. T. Oh, W. Han, U. C. Paek, and Y. Chung, "Discrimination of temperature and strain with a single FBG based on the birefringence effect," *Opt. Express*, vol. 12, no. 4, pp. 724–729, 2004.
11. X. Hu, C.-F. J. Pun, H.-Y. Tam, P. Mégret, and C. Caucheteur, "Highly reflective Bragg gratings in slightly etched step-index polymer optical fiber," *Opt. Express*, vol. 22, no. 15, pp. 18807–18817, 2014.
12. X. Hu et al., "Polarization effects in polymer FBGs: Study and use for transverse force sensing," *Opt. Express*, vol. 23, no. 4, pp. 4581–4590, 2015.
13. C. Caucheteur et al., "Transverse strain measurements using the birefringence effect in fiber Bragg gratings," *IEEE Photonics Technol. Lett.*, vol. 19, no. 13, pp. 966–968, 2007.
14. A. Stefani, S. Andresen, W. Yuan, and O. Bang, "Dynamic characterization of polymer optical fibers," *IEEE Sens. J.*, vol. 12, no. 10, pp. 3047–3053, 2012.
15. A. Stefani, S. Andresen, W. Yuan, N. Herholdt-Rasmussen, and O. Bang, "High sensitivity polymer optical fiber-Bragg-grating-based accelerometer," *IEEE Photonics Technol. Lett.*, vol. 24, no. 9, pp. 763–765, 2012.
16. A. Stefani et al., "Temperature compensated, humidity insensitive, high-Tg TOPAS FBGs for accelerometers and microphones," in *22nd International Conference on Optical Fiber Sensors*, Y. Liao et al., Eds. Bellingham: SPIE, 2012, vol. 8421, p. 84210Y.
17. I.-L. Bundalo, *Fibre Bragg Grating and Long Period Grating Sensors in Polymer Optical Fibres*. Denmark: Technical University of Denmark (DTU), 2017.
18. R. Ferreira, L. Bilro, C. Marques, R. Oliveira, and R. Nogueira, "Refractive index and viscosity: dual sensing with plastic fibre gratings," in *23rd International Conference on Optical Fibre Sensors*, J. M. López-Higuera et al., Eds. Bellingham: SPIE, 2014, vol. 9157, p. 915793.

19. C. A. F. Marques, L. Bilro, L. Kahn, R. A. Oliveira, D. J. Webb, and R. N. Nogueira, "Acousto-optic effect in microstructured polymer fiber bragg gratings: Simulation and experimental overview," *J. Light. Technol.*, vol. 31, no. 10, pp. 1551–1558, 2013.

20. D. Gallego and H. Lamela, "High-sensitivity ultrasound interferometric single-mode polymer optical fiber sensors for biomedical applications," *Opt. Lett.*, vol. 34, no. 12, pp. 1807–1809, 2009.

21. C. Broadway et al., "Fabry-Perot micro-structured polymer optical fibre sensors for opto-acoustic endoscopy," in *Biophotonics South America*, C. Kurachi, K. Svanberg, B. J. Tromberg and V. S. Bagnato, Eds. Bellingham: SPIE, 2015, vol. 9531, p. 953116.

22. C. Broadway et al., "Microstructured polymer optical fibre sensors for opto-acoustic endoscopy," in *Micro-Structured and Specialty Optical Fibres IV*, K. Kalli and A. Mendez, Eds. Bellingham: SPIE, 2016, vol. 9886, p. 98860S.

23. D. Gallego and H. Lamela, "Microstructured polymer optical fiber sensors for optoacoustic endoscopy," in *Photons Plus Ultrasound: Imaging and Sensing*, A. A. Oraevsky and L. V. Wang, Eds. Bellingham: SPIE, 2017, vol. 10064, p. 1006412.

24. C. Broadway et al., "A compact polymer optical fibre ultrasound detector," in *Photons Plus Ultrasound: Imaging and Sensing*, A. A. Oraevsky and L. V. Wang, Eds. Bellingham: SPIE, 2016, vol. 9708, p. 970813.

25. M. F. Ashby, *Materials Selection Mechanical Design*, 2nd ed. New Delhi: Pergamon Press Ltd, 1999.

26. D. Song, Z. Wei, Z. Chen, and H. Cui, "Liquid-level sensor using a fiber Bragg grating and carbon fiber composite diaphragm," *Opt. Eng.*, vol. 50, no. 1, p. 014401, 2011.

27. C. A. F. Marques, G. Peng, and D. J. Webb, "Highly sensitive liquid level monitoring system utilizing polymer fiber Bragg gratings," *Opt. Express*, vol. 23, no. 5, pp. 6058–6072, 2015.

28. G. Rajan, B. Liu, Y. Luo, E. Ambikairajah, and G.-D. Peng, "High sensitivity force and pressure measurements using etched singlemode polymer fiber Bragg gratings," *IEEE Sens. J.*, vol. 13, no. 5, pp. 1794–1800, 2013.

29. G. Rajan, K. Bhowmik, J. Xi, and G. D. Peng, "Etched polymer fibre Bragg gratings and their biomedical sensing applications," *Sensors*, vol. 17, no. 10, p. 2336, 2017.

30. I.-L. Bundalo, R. Lwin, S. Leon-Saval, and A. Argyros, "All-plastic fiber-based pressure sensor," *Appl. Opt.*, vol. 55, no. 4, pp. 811–816, 2016.

31. B. Schyrr et al., "Development of a polymer optical fiber pH sensor for on-body monitoring application," *Sensors Actuators B Chem.*, vol. 194, pp. 238–248, 2014.

32. X. H. Wang and L. L. Yang, "Fluorescence pH probe based on microstructured polymer optical fiber," *Opt. Express*, vol. 15, no. 25, pp. 16479–16483, 2007.

33. A. Richter, G. Paschew, S. Klatt, J. Lienig, K. F. Arndt, and H. J. P. Adler, "Review on hydrogel-based pH sensors and microsensors," *Sensors*, vol. 8, no. 1, pp. 561–581, 2008.

34. A. Azhari et al., "Development of fiber Bragg grating pH sensors for harsh environments," in *Fiber Optic Sensors and Applications XV*, A. Mendez, C. S. Baldwin and H. H. Du, Eds. Bellingham: SPIE, 2018, vol. 10654, p. 106540P.

35. T. X. Wang, Y. H. Luo, G. D. Peng, and Q. J. Zhang, "High-sensitivity stress sensor based on Bragg grating in BDK-doped photosensitive polymer optical fiber," in *Third Asia Pacific Optical Sensors Conference*, J. Canning and G.-D. Peng, Eds. Bellingham: SPIE, 2012, vol. 8351, p. 83510M.

36. X. Cheng, J. Bonefacino, B. O. Juan, and H. Y. Tam, "All-polymer fiber-optic pH sensor," *Opt. Express*, vol. 26, no. 11, pp. 14610–14616, 2018.

37. J. Janting, J. Pedersen, G. Woyessa, K. Nielsen, and O. Bang, "Small and robust all-polymer fiber Bragg grating based pH sensor," *J. Light wave Technol.*, vol. 37, no. 18, pp. 4480–4486, 2019.

38. A. Pospori et al., "Annealing effects on strain and stress sensitivity of polymer optical fibre based sensors," in *Micro-Structured and Specialty Optical Fibres IV*, K. Kalli and A. Mendez, Eds. Bellingham: SPIE, 2016, vol. 9886, p. 98860V.

39. W. Zhang and D. J. Webb, "Humidity responsivity of poly(methyl methacrylate)-based optical fiber Bragg grating sensors," *Opt. Lett.*, vol. 39, no. 10, pp. 3026–3029, 2014.

40. W. Yuan et al., "Humidity insensitive TOPAS polymer fiber Bragg grating sensor," *Opt. Express*, vol. 19, no. 20, pp. 19731–19739, 2011.

41. J. Bonefacino et al., "Ultra-fast polymer optical fibre Bragg grating inscription for medical devices," *Light Sci. Appl.*, vol. 7, no. 3, p. 17161, 2018.

42. K. Krebber, P. Lenke, S. Liehr, J. Witt, and M. Schukar, "Smart technical textiles with integrated POF sensors," in *Smart Sensor Phenomena, Technology, Networks, and Systems*, W. Ecke, K.J. Peters, N. G. Meyendorf, Eds. Bellingham: SPIE, 2008, vol. 6933, p. 69330V.

43. B. C. Yao et al., "Graphene-based D-shaped polymer FBG for highly sensitive erythrocyte detection," *IEEE Photonics Technol. Lett.*, vol. 27, no. 22, pp. 2399–2402, 2015.

44. D. Vilarinho et al., "POFBG-embedded cork insole for plantar pressure monitoring," *Sensors*, vol. 17, no. 12, p. 2924, 2017.

45. A. Leal-Junior et al., "Fiber Bragg gratings in CYTOP fibers embedded in a 3D-printed flexible support for assessment of human–robot interaction forces," *Materials*, vol. 11, no. 11, p. 2305, 2018.

46. A. Grillet et al., "Optical fiber sensors embedded into medical textiles for healthcare monitoring," *IEEE Sens. J.*, vol. 8, no. 7, pp. 1215–1222, 2008.

47. C. C. Ye et al., "Applications of polymer optical fibre grating sensors to condition monitoring of textiles," *J. Phys. Conf. Ser.*, vol. 178, no. 1, p. 012020, 2009.

48. X. Chen et al., "Photonic skin for pressure and strain sensing," in *Optical Sensors and Detection*, A. G. Mignani and C. A. van Hoof, Eds. Bellingham: SPIE, 2010, vol. 7726, p. 772604.

49. M. F. Domingues et al., "Insole optical fiber Bragg grating sensors network for dynamic vertical force monitoring," *J. Biomed. Opt.*, vol. 22, no. 9, p. 091507, 2017.

50. G. Rajan, M. Ramakrishnan, Y. Semenova, E. Ambikairajah, G. Farrell, and G.-D. Peng, "Experimental study and analysis of a polymer fiber Bragg grating embedded in a composite material," *J. Light. Technol.*, vol. 32, no. 9, pp. 1726–1733, 2014.

51. M. G. Zubel, K. Sugden, D. J. Webb, D. Sáez-Rodríguez, K. Nielsen, and O. Bang, "Embedding silica and polymer fibre Bragg gratings (FBG) in plastic 3D-printed sensing patches," in *Micro-Structured and Specialty Optical Fibres IV*, K. Kalli and A. Mendez, Eds. Bellingham: SPIE, 2016, vol. 9886, p. 98860N.

52. F. Berghmans et al., "Poly(D,L-lactic acid) (PDLLA) biodegradable and biocompatible polymer optical fiber," *J. Light. Technol.*, vol. 37, no. 9, pp. 1916–1923, 2019.

53. G. Emiliyanov et al., "Localized biosensing with Topas microstructured polymer optical fiber," *Opt. Lett.*, vol. 32, no. 5, pp. 460–462, 2007.

54. J. B. Jensen, P. E. Hoiby, and L. H. Pedersen, "Selective detection of antibodies in microstructured polymer optical fibers," *Opt. Express*, vol. 13, no. 15, pp. 5883–5889, 2005.

55. A. Gutés, C. Carraro, and R. Maboudian, "Single-layer CVD-grown graphene decorated with metal nanoparticles as a promising biosensing platform," *Biosens. Bioelectron.*, vol. 33, no. 1, pp. 56–59, 2012.

56. K. Jansen, G. Van Soest, and A. F. W. van der Steen, "Intravascular photoacoustic imaging: A new tool for vulnerable plaque identification," *Ultrasound Med. Biol.*, vol. 40, no. 6, pp. 1037–1048, 2014.

57. D. J. Webb et al., "Miniature fibre optic ultrasonic probe," in *International Symposium on Optical Science, Engineering, and Instrumentation*, 1996, vol. 2639, pp. 76–80.

58. A. D. Kersey et al., "Fiber grating sensors," *J. Light. Technol.*, vol. 15, no. 8, pp. 1442–1463, 1997.
59. M. G. Xu, L. Reekie, J. P. Dakin, and J.-L. Archambault, "Discrimination between strain and temperature effects using dual-wavelength fibre grating sensors," *Electron. Lett.*, vol. 30, no. 13, pp. 1085–1087, 1994.
60. G. D. Peng, Z. Xiong, and P. L. Chu, "Photosensitivity and gratings in dye-doped polymer optical fibers," *Opt. Fiber Technol.*, vol. 5, no. 2, pp. 242–251, 1999.
61. Z. Xiong, G. D. Peng, B. Wu, and P. L. Chu, "Highly tunable Bragg gratings in single-mode polymer optical fibers," *IEEE Photonics Technol. Lett.*, vol. 11, no. 3, pp. 352–354, 1999.
62. H. Y. Liu, G. D. Peng, and P. L. Chu, "Thermal tuning of polymer optical fiber Bragg gratings," *IEEE Photonics Technol. Lett.*, vol. 13, no. 8, pp. 824–826, 2001.
63. H. B. Liu, H. Y. Liu, G. D. Peng, and P. L. Chu, "Strain and temperature sensor using a combination of polymer and silica fibre Bragg gratings," *Opt. Commun.*, vol. 219, no. 1–6, pp. 139–142, 2003.
64. K. Bhowmik et al., "Intrinsic high-sensitivity sensors based on etched single-mode polymer optical fibers," *IEEE Photonics Technol. Lett.*, vol. 27, no. 6, pp. 604–607, 2015.
65. K. Bhowmik et al., "High intrinsic sensitivity etched polymer fiber Bragg grating pair for simultaneous strain and temperature measurements," *IEEE Sens. J.*, vol. 16, no. 8, pp. 2453–2459, 2016.
66. A. Abang, D. Saez-Rodriguez, K. Nielsen, O. Bang, and D. J. Webb, "Connectorisation of fibre Bragg grating sensors recorded in microstructured polymer optical fibre," in *Fifth European Workshop on Optical Fibre Sensors (EWOFS'2013)*, L. R. Jaroszewicz, Ed. Bellingham: SPIE, 2013, vol. 8794, p. 87943Q.
67. R. Lwin and A. Argyros, "Connecting microstructured polymer optical fibres to the world," in *18th International Conference on Plastic Optical Fibers*, 2009, vol. 18, p. Poster 7.
68. R. Oliveira, L. Bilro, and R. Nogueira, "Fabry-Perot cavities based on photopolymerizable resins for sensing applications," *Opt. Mater. Express*, vol. 8, no. 8, pp. 899–902, 2018.
69. R. Oliveira, L. Bilro, R. Nogueira, and A. M. Rocha, "Adhesive based Fabry-Pérot hydrostatic pressure sensor with improved and controlled sensitivity," *J. Light. Technol.*, vol. 37, no. 9, pp. 1909–1915, 2019.
70. R. Oliveira, L. Bilro, and R. Nogueira, "Strain and temperature detection through PFBG and resin based FP cavities," in *26th International Conference on Optical Fiber Sensors*, L. Thévenaz et al., Eds. Washington, DC: OSA, 2018, p. WF78.
71. R. Oliveira, T. H. R. Marques, L. Bilro, R. Nogueira, and C. M. B. Cordeiro, "Multiparameter POF sensing based on multimode interference and fiber Bragg grating," *J. Light. Technol.*, vol. 35, no. 1, pp. 3–9, 2017.
72. F. M. Haran, J. K. Rew, and P. D. Foote, "A strain-isolated fibre Bragg grating sensor for temperature compensation of fibre Bragg grating strain sensors," *Meas. Sci. Technol.*, vol. 9, pp. 1163–1166, 1998.
73. W. Yuan, A. Stefani, and O. Bang, "Tunable polymer fiber Bragg grating (FBG) inscription: Fabrication of dual-FBG temperature compensated polymer optical fiber strain sensors," *Photonics Technol. Lett.*, vol. 24, no. 5, pp. 401–403, 2012.
74. W. Qiu, X. Cheng, Y. Luo, Q. Zhang, and B. Zhu, "Simultaneous measurement of temperature and strain using a single Bragg grating in a few-mode polymer optical fiber," *J. Light. Technol.*, vol. 31, no. 14, pp. 2419–2425, 2013.
75. R. Min, B. Ortega, and C. A. F. Marques, "Fabrication of tunable chirped mPOF Bragg gratings using a uniform phase mask," *Opt. Express*, vol. 26, no. 4, pp. 4411–4420, 2018.

76. R. Min et al., "Microstructured PMMA POF chirped Bragg gratings for strain sensing," *Opt. Fiber Technol.*, vol. 45, pp. 330–335, 2018.

77. N. Harbach, *Fiber Bragg Gratings in Polymer Optical Fibers*. Lausanne: École Polytechnique Fédérale de Lausanne, 2008.

78. G. N. Harbach, H. G. Limberger, and R. P. Salathé, "Influence of humidity and temperature on polymer optical fiber Bragg gratings," in *Advanced Photonics & Renewable Energy*, 2010, p. BTuB2.

79. C. Markos, A. Stefani, K. Nielsen, H. K. Rasmussen, W. Yuan, and O. Bang, "High-Tg TOPAS microstructured polymer optical fiber for fiber Bragg grating strain sensing at 110 degrees," *Opt. Express*, vol. 21, no. 4, pp. 4758–4765, 2013.

80. G. Woyessa, A. Fasano, A. Stefani, C. Markos, H. K. Rasmussen, and O. Bang, "Single mode step-index polymer optical fiber for humidity insensitive high temperature fiber Bragg grating sensors," *Opt. Express*, vol. 24, no. 2, pp. 1253–1260, 2016.

81. G. Woyessa, A. Fasano, C. Markos, A. Stefani, H. K. Rasmussen, and O. Bang, "Zeonex microstructured polymer optical fiber: Fabrication friendly fibers for high temperature and humidity insensitive Bragg grating sensing," *Opt. Mater. Express*, vol. 7, no. 1, pp. 286–295, 2017.

82. R. Min, B. Ortega, A. Leal-Junior, and C. Marques, "Fabrication and characterization of Bragg grating in CYTOP POF at 600-nm wavelength," *IEEE Sensors Lett.*, vol. 2, no. 3, 2018.

83. H. Sun, X. Zhang, L. Yuan, L. Zhou, X. Qiao, and M. Hu, "An optical fiber Fabry-Perot interferometer sensor for simultaneous measurement of relative humidity and temperature," *IEEE Sens. J.*, vol. 15, no. 5, pp. 2891–2897, 2015.

84. C.-T. Ma, Y.-W. Chang, Y.-J. Yang, and C.-L. Lee, "A dual-polymer fiber fizeau interferometer for simultaneous measurement of relative humidity and temperature," *Sensors*, vol. 17, no. 11, p. 2659, 2017.

85. C. Massaroni, M. Caponero, R. D'Amato, D. Lo Presti, and E. Schena, "Fiber bragg grating measuring system for simultaneous monitoring of temperature and humidity in mechanical ventilation," *Sensors*, vol. 17, no. 4, p. 749, 2017.

86. M. Ams et al., "Fibre optic temperature and humidity sensors for harsh wastewater environments," in *Eleventh International Conference on Sensing Technology (ICST)*, 2017, pp. 1–3.

87. C. Zhang, W. Zhang, D. J. Webb, and G.-D. Peng, "Optical fibre temperature and humidity sensor," *Electron. Lett.*, vol. 46, no. 9, pp. 643–644, 2010.

88. G. Woyessa et al., "Zeonex-PMMA microstructured polymer optical FBGs for simultaneous humidity and temperature sensing," *Opt. Lett.*, vol. 42, no. 6, pp. 1161–1164, 2017.

89. R. Oliveira, L. Bilro, and R. Nogueira, "Simultaneous measurement of temperature and humidity using PFBG and Fabry-Perot cavity," in *26th International Conference on Plastic Optical Fibres*, 2017, p. Paper 83.

90. R. Oliveira, L. Bilro, T. H. R. Marques, C. M. B. Cordeiro, and R. Nogueira, "Simultaneous detection of humidity and temperature through an adhesive based Fabry – Pérot cavity combined with polymer fiber Bragg grating," *Opt. Lasers Eng.*, vol. 114, pp. 37–43, 2019.

91. W. L. Chen, K. R. Shull, T. Papatheodorou, D. A. Styrkas, and J. L. Keddie, "Equilibrium swelling of hydrophilic polyacrylates in humid environments," *Macromolecules*, vol. 32, no. 1, pp. 136–144, 1999.

92. W. Zhang, D. J. Webb, and G.-D. Peng, "Investigation into time response of polymer fiber Bragg grating based humidity sensors," *J. Light. Technol.*, vol. 30, no. 8, pp. 1090–1096, 2012.

Index